Sustainable Development Goals Series

The **Sustainable Development Goals Series** is Springer Nature's inaugural cross-imprint book series that addresses and supports the United Nations' seventeen Sustainable Development Goals. The series fosters comprehensive research focused on these global targets and endeavours to address some of society's greatest grand challenges. The SDGs are inherently multidisciplinary, and they bring people working across different fields together and working towards a common goal. In this spirit, the Sustainable Development Goals series is the first at Springer Nature to publish books under both the Springer and Palgrave Macmillan imprints, bringing the strengths of our imprints together.

The Sustainable Development Goals Series is organized into eighteen subseries: one subseries based around each of the seventeen respective Sustainable Development Goals, and an eighteenth subseries, "Connecting the Goals", which serves as a home for volumes addressing multiple goals or studying the SDGs as a whole. Each subseries is guided by an expert Subseries Advisor with years or decades of experience studying and addressing core components of their respective Goal.

The SDG Series has a remit as broad as the SDGs themselves, and contributions are welcome from scientists, academics, policymakers, and researchers working in fields related to any of the seventeen goals. If you are interested in contributing a monograph or curated volume to the series, please contact the Publishers: Zachary Romano [Springer; zachary.romano@springer.com] and Rachael Ballard [Palgrave Macmillan; rachael.ballard@palgrave.com].

Nor Aida Abdul Rahman ·
Kamran Mahroof · Juliater Simarmata
Editors

Supply Chain in Humanitarian Operations

Managing Disruption Through Intervention and Innovation

Springer

Editors
Nor Aida Abdul Rahman
Universiti Kuala Lumpur
Malaysian Institute of Aviation Technology
Subang, Selangor, Malaysia

Kamran Mahroof
Faculty of Management, Law & Social Sciences
University of Bradford
Bradford, UK

Juliater Simarmata
Institute of Transportation and Logistics
Jakarta, Indonesia

ISSN 2523-3084 ISSN 2523-3092 (electronic)
Sustainable Development Goals Series
ISBN 978-981-96-4425-4 ISBN 978-981-96-4426-1 (eBook)
https://doi.org/10.1007/978-981-96-4426-1

This Springer imprint is published by the registered company Springer Nature Singapore Pte Ltd.
The registered company address is: 152 Beach Road, #21-01/04 Gateway East, Singapore 189721, Singapore

To all wonderful readers, contributors, colleagues, spouse, children, parents and siblings:
Ir. Jailani, Muda, Dr.Hj. Abdul Rahman, Hjh. Radziah, Hjh Halimah Tahar, Danial Adam, Balqis Adriana, Dani Alif, Hud Aqil, Noah Eusuff,, Aziah, Azman and Azam.

—Nor Aida Abdul Rahman—NAAR

To my family, colleagues, and readers—your interest, support, guidance, and curiosity have inspired every page of this book

—Kamran Mahroof

To my family who stood by me, my colleagues who encouraged me,
and the readers who make the journey worthwhile

—Juliater Simarmata

Preface

Humanitarian logistics and supply chain management play a pivotal role in ensuring effective response and relief efforts during times of crises. The main aim of humanitarian logistics and supply chain management is to ensure the effectiveness and efficiency of the humanitarian planning process, implementing, and controlling the efficient and timely flow and storage of essential goods, services, and critical information from the point of origin to the point of consumption in a complex and often challenging humanitarian context, ultimately aiming to alleviate suffering and save lives. At the same time, technology and appropriate systems are also crucial for addressing the urgent needs of communities affected by disasters, conflicts, and emergencies. In recent crisis era, the significance of humanitarian logistics and supply chain management cannot be overstated especially in the area of food supply, medical relief, healthcare service, counselling, and other service. These are the main instrumental systems that help affected populations receive food, water, medical supplies, shelter, and other vital aid in a timely and organized manner.

A well-organized and effective supply chain infrastructure is necessary due to the complexity of crises and the need for quick responses. Governmental and non-governmental humanitarian organizations have to deal with issues like erratic weather patterns, inadequate infrastructure, security threats, and culturally heterogeneous environments. The ability to manage and optimize the logistics and supply chain processes becomes a determining factor in the success of relief operations and the overall impact on affected communities.

This book highlights the vital role that humanitarian logistics and supply chain management play in delivering aid during times of crisis with recent technology application. As we delve deeper into this field, we will explore the key components, challenges, and innovations that contribute to the effectiveness of these systems in providing timely and targeted assistance during times of adversity.

The development of technology leads us to believe that it possesses a certain astuteness that influences the way logistical and humanitarian efforts are carried out. Nowadays, practically every area of humanitarian activities benefits from the expanded use of technology. This enriches the description, explanation, assessment, and use of technology in the academic and research domains. Technology plays important roles in making humanitarian activities flourish and sustain. Indeed, as the global disaster report from 2013 has made clear, it is critical to consider the significant influence that technological advancements have had on humanitarian efforts, including the novel and inventive ways in which humanitarians have used technology and the potential risks and opportunities that may arise from these advancements. The responsible use of technology provides tangible methods to enhance the efficacy, efficiency, and accountability of humanitarian aid efforts. This strategic implementation of technology can, as a consequence, directly mitigate vulnerabilities within affected populations and bolster their resilience against future crises. Moreover, it enables aid organizations to reach more beneficiaries, optimize resource allocation, and improve real-time decision-making in complex emergency situations, ultimately leading to more sustainable and impactful humanitarian outcomes.

Humanitarian logistics and supply chain activities also contribute to several Sustainable Development Goals (SDGs), primarily those related to poverty alleviation, health, and well-being, as well as disaster resilience. For instance, SDG 1 relates with No Poverty. Humanitarian logistics aims to provide essential goods and services to those affected by crises, helping to alleviate poverty in emergency situations. While SDG goal no 2 relates with Zero Hunger. Humanitarian logistics and supply chain play a crucial role in delivering food aid and ensuring access to nutrition in humanitarian emergencies. SDG 3 goal is on Good Health and Well-being. In this perspective, humanitarian logistics contribute to the provision of medical supplies, equipment, and healthcare services in crisis-affected areas, promoting health and well-being. Additionally, humanitarian logistics and supply chain also connected to other SDG goals including SDG 6 on clean water, SDG 9 on industry, innovation and infrastructure, SDG 11 on sustainable cities, SDG 13 on climate action, as well as SDG 17 on partnership for the goals. In this book, the reader will be exposed with all these SDG goals and their relation to technology application in humanitarian logistics and supply chain activities. Issue related to collaboration between government, NGOs, and private sector will also be discussed.

On this note, this book aims to focus on various trends and technologies in humanitarian logistics and supply chain, and its relation to SDG goals' contributions. This book which contains of fourteen (14) chapters covers issues of technology revolution in the field of humanitarian logistics and supply chain, including its emerging trends. This book is packed with detailed knowledge on both theories and applications of the humanitarian logistics and supply chain components with recent technology application to ensure effective and efficient delivery of aid during emergencies or crises. These include needs assessment, strategic planning, information management, coordination and collaboration, infrastructure and facilities, transportation, custom issue, training and security management.

This book gives the reader a deeper, up-to-date insight into technology and emerging trends in the context of humanitarian logistics and supply chain. This book not only offers a blend of comprehensive and extensive high-quality research outputs from highly reputed authors and editors, but also their commitment to advancing the frontiers of knowledge through meticulous analysis and innovative thinking.

"Supply Chain in Humanitarian Operations: Managing Disruption Through Intervention and Innovation" offers readers a comprehensive and cutting-edge exploration of technology's role in revolutionizing humanitarian logistics and supply chain management. This seminal work brings together contributions from world-renowned experts, presenting a synthesis of rigorous academic research and real-world applications.

The book addresses three pivotal objectives:

a. Technology Integration and Optimization: It delves deep into how emerging technologies can dramatically enhance the efficiency and effectiveness of humanitarian logistics and supply chain operations. From blockchain for transparent aid distribution to AI-powered predictive analytics for disaster response, the book examines transformative solutions that are reshaping the field.
b. Building Resilience in Crisis: In an era of increasing global instability, this volume investigates how technological advancements can empower humanitarian organizations to develop robust strategies and systems. It explores cutting-edge approaches to creating adaptive, resilient supply chains capable of withstanding and rapidly recovering from disruptions, whether they be natural disasters, conflicts, or pandemics.
c. Sustainable Development Goals (SDGs) in Humanitarian Logistics and Fostering Multi-Stakeholder Collaboration: The book provides a critical analysis of how the humanitarian sector can align its operations with the United Nations' SDGs. It offers innovative frameworks for integrating sustainability into every aspect of humanitarian supply chains, from eco-friendly packaging to carbon-neutral transportation methods. It presents strategies for leveraging digital platforms to create synergies between NGOs, governments, private sector partners, and affected communities.

This groundbreaking work seamlessly blends theoretical rigor with practical insights, making it an invaluable resource for a wide audience. While primarily targeted at doctoral candidates, postgraduate and graduate students in fields such as humanitarian logistics, supply chain management, disaster response, and international development, its accessible yet profound content also makes it indispensable for:

- Policymakers and government officials involved in disaster preparedness and response;
- NGO leaders and humanitarian aid professionals seeking to optimize their operations;
- Corporate executives looking to engage in meaningful CSR initiatives or enter the humanitarian sector;
- Researchers and academics pushing the boundaries of knowledge in this critical field;
- Technology innovators developing solutions for global challenges.

By bridging the gap between academic theory and real-world application, *Supply Chain in Humanitarian Operations* not only advances scholarly discourse but also provides actionable strategies for improving humanitarian responses worldwide. It stands as a testament to the power of innovation in alleviating human suffering and building a more resilient global community.

Chapter 1 from the editors, Rahman et al highlight the current update of the research on humanitarian logistics and supply chain with providing summary of metadata for 243 articles on technology and humanitarian supply chain from Scopus database. It is followed by Zahari et al. in Chap. 2 describing the optimization of humanitarian logistics. Chapter 3 by Osman and Ustadi provides thematic review on critical success factors in humanitarian logistics. Othman and Ahmad in Chap. 4 discuss the innovation adoption in humanitarian logistics. Mayor-Vitoria in Chap. 5 focuses on the emerging role of technology in mitigating risk in humanitarian supply chain. Chapter 6 by Hamid and Alemu discusses on the use of AI in assisting humanitarian efforts. While in Chap. 7 by Khairuddin and Rahman, the authors provide IOT-enabled solution for distribution centers as part of the humanitarian effort activities. In Chap. 8, Ephraim discusses on how blockchain technology manages supply chain risk. In the ninth chapter, Baeda provide discussion on the role of technology in mitigating risk in humanitarian supply chain with presenting Indonesian Archipelago as a case study. In the tenth chapter, Hamid et al discuss on AI-powered humanities assistance with exploring the key opportunities and the potential risk. In Chap. 11, Jailani et al. highlight the integration of technology and medical services in humanitarian operation from multiple case study approaches. Hussain and Amathalai in Chap. 12 explain the biometric technology application in humanitarian sector. In Chap. 13, Rumani and Nasution discuss the AI for optimizing humanitarian aid from air traffic management perspectives. Osman in the final chapter, Chap. 14, provides discussion on humanitarian logistics and the lesson learnt from recent pandemic COVID-19.

Subang, Malaysia — Nor Aida Abdul Rahman
Bradford, UK — Kamran Mahroof
Jakarta, Indonesia — Juliater Simarmata

Contents

Editors and Contributors

About the Editors

Dr. Nor Aida Abdul Rahman currently serves as an Associate Professor at Universiti Kuala Lumpur, Malaysian Institute of Aviation Technology (UniKL MIAT) in Subang, Selangor, Malaysia. Her work has been featured in prestigious journals such as *Industrial Marketing Management*, *Journal of Islamic Marketing*, *Journal of Humanitarian Logistics and Supply Chain Management*, and *Journal of Quality and Reliability Management*, among others. In addition to her role at UniKL MIAT, she holds visiting appointments as a Professor at ITL Trisakti in Jakarta, Indonesia, and as a Visiting Fellow at the UKM-MPOB Endowment Chair at the National University of Malaysia. Dr. Nor Aida has a distinguished record of scholarship, having authored or co-authored over 100 articles, book chapters, and research books. Her expertise also extends to the publishing realm, where she has contributed to the editorial teams of more than 10 books with Routledge and Springer. She earned her Ph.D. in Management (Supply Chain Management) from Brunel University, London, after completing her Master's and Bachelor's degrees in Business and Administration from the National University of Malaysia. Her dedication to the field is further demonstrated by her roles as an MQA panel auditor, a JAKIM halal-certified trainer, an HRDF certified trainer, and as a Board Member of Education at the Chartered Institute of Logistics and Transport Malaysia (CILTM).

Dr. Kamran Mahroof is an Associate Professor of Supply Chain Analytics and Programme Leader for the M.Sc. in the Applied Artificial Intelligence and Data Analytics programme, at the University of Bradford, School of Management. He has extensive industry experience having worked in various role for leading organisations, most recently for Morrisons Plc, where he led various productivity initiatives in Logistics. Kamran continues to hold a strong connection in several industries through his ongoing research. His industrial experience has helped shape his research interests, whereby he is actively researching in the areas of Supply Chain Management, Circular Economy, Artificial Intelligence (AI) and the Societal Impact of Emerging Technologies. He regularly publishes in high quality, high impact peer-reviewed journals, such as *International Journal of Production Economics*, *Journal of Business Research and Information Systems Frontiers*. Kamran makes regular appearances on national television, radio and in other media and has written several articles for the Press, along with other internationally renowned publications, such as for the World Economic Forum. He is also a Fellow of the Higher Education (FHEA).

Dr. Juliater Simarmata, SE, MM is a Lecturer and Researcher at the Trisakti Institute of Transportation and Logistics, Jakarta—Indonesia. He has several articles and books on artificial intelligence marketing, destination management organization, and many more published in national and international journals. His research interests include airline marketing, marketing strategy, and consumer behavior. Apart from being a lecturer, he is active as a training consultant and speaker in the fields of Sales/Marketing, Business Management, and Entrepreneurship in various companies in Indonesia.

Contributors

Zulkarnian Ahmad Faculty of Business and Management, Akademi Laut Malaysia, Melaka, Malaysia

Abebe Ejigu Alemu Department of Logistics Management, International Maritime College Oman, National University of Science and Technology, Muscat, Oman

Ramesh Babu a/l Amathalai Faculty of Maritime Management, Akademi Laut Malaysia, Melaka, Malaysia

Achmad Yasir Baeda Marine Disaster and Weather Anomalies Research Laboratory (MDWA-RL), Faculty of Engineering, Universitas Hasanuddin, Makassar, Indonesia

Abdelsalam Adam Hamid College of Business Administration, A'Sharqiyah University, Ibra, Oman

Mohd Azizi bin Hamid Centre for Academic Quality, Akademi Laut Malaysia, Melaka, Malaysia

Nur Hidayah Azhar National Defence University of Malaysia, Kuala Lumpur, Malaysia

Nurul Ain Safwah binti Ekey Hussain Faculty of Maritime Management, Akademi Laut Malaysia, Melaka, Malaysia

Muhammad Danial Adam Jailani USM-KLE International Medical School, Belagavi, Karnataka, India

Nur Balqis Adriana Jailani Manipal University College Malaysia, Melaka, Malaysia

Muhamad Munzir Khairuddin Universiti Kuala Lumpur, Malaysian Institutc of Aviation Technology, Subang, Selangor, Malaysia

Kamran Mahroof Faculty of Management, Law & Social Sciences, Unversity of Bradford, Bradford, UK

Fernando Mayor-Vitoria Universidad Internacional de Valencia—VIU, Valencia, Spain

Ruzaidin Mohammed Zain Strategic and Security Studies Institute, HADR Research Center, National Defence University of Malaysia, Kuala Lumpur, Malaysia

Najuwa Mohd Nasir Faculty of Business and Communication, Universiti Malaysia Perlis, Arau, Malaysia

Haliza Mohd Zahari Strategic and Security Studies Institute, HADR Research Center, National Defence University of Malaysia, Kuala Lumpur, Malaysia

Darmeli Nasution Universitas Pembangunan Pancabudi, Medan, Indonesia

Sharina Osman Business School, Universiti Kuala Lumpur, Kuala Lumpur, Malaysia

Aisyah Othman Faculty of Business and Management, Akademi Laut Malaysia, Melaka, Malaysia

Suzari Abdul Rahim Graduate School of Business, Universiti Sains Malaysia, Pulau Pinang, Malaysia

Nadia Harnisa binti Abdul Rahman Faculty of Maritime Management, Akademi Laut Malaysia, Melaka, Malaysia

Nor Aida Abdul Rahman Universiti Kuala Lumpur, Malaysian Institute of Aviation Technology, Subang, Selangor, Malaysia

Daniel D. Rumani Akademi Penerbang Indonesia, Banyuwangi, Indonesia

Jeremy Ephraim Simarmata Department of Business Management, Trisakti Institute of Transportation and Logistics, East Jakarta, Indonesia

Juliater Simarmata Institute of Transportation and Logistics, Jakarta, Indonesia

Shahidah binti Ahmad Suhaimi Faculty of Maritime Management, Akademi Laut Malaysia, Melaka, Malaysia

Muna Norkhairunnisak Ustadi Business School, Universiti Kuala Lumpur, Kuala Lumpur, Malaysia

List of Figures

List of Tables

Part I
Introduction to Humanitarian Logistics and Supply Chain

Research on Humanitarian Logistics and Supply Chain Management

1

Nor Aida Abdul Rahman, Kamran Mahroof, and Juliater Simarmata

Abstract

This chapter explores the critical field of humanitarian logistics and supply chain management, focusing on its unique challenge in managing disruptions, innovative solutions, and theoretical underpinnings. As humanitarian crises become increasingly complex and frequent, the need for resilient and adaptive supply chains has never been more pressing. The research explores various types of disruptions affecting humanitarian supply chains, from natural disasters to man-made crises, and analyzes effective intervention strategies to mitigate their impact. The chapter highlights innovative approaches leveraging cutting-edge technologies such as artificial intelligence, blockchain, and additive manufacturing to enhance the agility and effectiveness of humanitarian supply chains. A comprehensive review of relevant past provides a theoretical foundation for understanding disruption management in humanitarian contexts. By synthesizing current knowledge and identifying future research directions, this work contributes to the ongoing development of more efficient and effective humanitarian logistics systems, ultimately aiming to improve the delivery of aid to those in need during crises.

Keywords

Humanitarian logistics · Supply chain management · Disaster response · Emergency preparedness · Technological innovation · Stakeholder coordination

N. A. A. Rahman (✉)
Universiti Kuala Lumpur, Malaysian Institute of Aviation Technology,
Subang, Selangor, Malaysia
e-mail: noraida@unikl.edu.my

K. Mahroof
Faculty of Management, Law & Social Sciences,
Unversity of Bradford, Bradford, UK
e-mail: K.Mahroof@bradford.ac.uk

J. Simarmata
Institute of Transportation and Logistics,
Jakarta, Indonesia
e-mail: juliaters@itltrisakti.ac.id

1.1 Introduction

Humanitarian supply chains play a critical role in providing essential aid and resources to populations affected by disasters, conflicts, and other emergencies. These complex networks involve numerous stakeholders, including non-governmental organizations, governments, donors, and local communities, working together to deliver vital assistance in often challenging environments (Van Wassenhove 2006). The effectiveness of humanitarian operations hinges on the

N. A. A. Rahman et al. (eds.), *Supply Chain in Humanitarian Operations*, Sustainable Development Goals Series,
https://doi.org/10.1007/978-981-96-4426-1_1

ability of supply chains to rapidly mobilize and distribute resources where they are most needed. In recent years, humanitarian supply chains have faced increasing pressure due to the growing frequency and severity of natural disasters, protracted conflicts, and global health crises. The COVID-19 pandemic, in particular, has exposed vulnerabilities in these systems and highlighted the need for greater resilience and adaptability. As Kovács and Falagara Sigala (2021) note, the pandemic caused unprecedented disruptions to global supply chains, affecting the transportation and provision of critical humanitarian commodities, including medical supplies, food, and non-food items.

Disruptions in humanitarian supply chains can take various forms and stem from multiple sources. Natural disasters, such as earthquakes, hurricanes, and floods, can damage infrastructure and impede access to affected areas (Day 2014). Political instability and conflicts may restrict movement and create security risks for aid workers. Moreover, economic crises can lead to shortages of essential goods and price volatility (Davarzani et al. 2015). The complex system of supply chain network and interconnected nature of global supply chains makes them vulnerable to cascading effects when disruptions occur in one part of the network (Fujita and Hamaguchi 2016). The impact of these disruptions can be severe, potentially leading to delays in aid delivery, increased costs, and ultimately, a reduced ability to meet the needs of vulnerable populations. For instance, during the acute phase of the COVID-19 pandemic, many humanitarian organizations faced challenges in procuring and distributing personal protective equipment (PPE) due to global shortages and export restrictions (Travers and Lowe 2021). This not only hindered their ability to protect their staff but also limited their capacity to continue essential operations safely. In response to these challenges, humanitarian organizations have been compelled to adapt their supply chain strategies and explore innovative approaches to enhance resilience. One of the example on the approach has been the localization of supply chains, as observed during the COVID-19 response. Humanitarian actors increasingly turned to local suppliers and manufacturers to meet urgent needs, particularly for items like PPE and hand sanitizer (McDougall and Davis 2024). This shift not only helped overcome global supply chain disruptions but also contributed to local economic recovery and capacity building.

The adoption of new technologies has also emerged as a key strategy for improving the efficiency and resilience of humanitarian supply chains. Digital platforms, blockchain, and data analytics are being leveraged to enhance transparency, traceability, and coordination among different actors in the supply chain (Sarkis 2021). For example, some organizations have implemented electronic cash transfer programs to provide assistance more efficiently and safely during the pandemic (Travers 2024).

Collaboration and knowledge sharing among humanitarian organizations have proven crucial in navigating supply chain disruptions. The establishment of common supplier databases, joint procurement initiatives, and the sharing of best practices have helped organizations leverage collective resources and expertise to overcome challenges (Herlin and Pazirandeh 2015). The COVID-19 Supply Chain System, launched by the World Health Organization and partners, exemplifies such collaborative efforts to ensure equitable access to essential supplies during the pandemic (Travers and Lowe 2021).

As the humanitarian sector continues to grapple with supply chain disruptions, several key areas warrant further research and attention. Understanding the types of disruptions that affect humanitarian supply chains and their specific impacts is crucial for developing targeted mitigation strategies. With that, this study aims to shed light on recent research related to supply chain disruption in humanitarian context and what are the key technologies discussed in previous research in supply chain humanitarian effort.

Case studies of successful responses to disruptions offer valuable insights for future preparedness. For instance, the COVID-19 pandemic and disaster response efforts have demonstrated innovative adaptations and technological solutions. These examples showcase how organizations have leveraged local resources, implemented flexible procurement, and utilized digital tools to overcome supply chain challenges (Saiah et al. 2023). Looking ahead, emerging trends and future research directions in humanitarian supply chain management are likely to focus on several key areas. These may include the further integration of advanced technologies like artificial intelligence and the Internet of Things to enhance supply chain visibility and decision-making (Kapoor et al. 2021). The development of more robust risk assessment and management tools tailored to the humanitarian context will be essential. Additionally, exploring sustainable and environmentally friendly supply chain practices that align with broader humanitarian goals will be increasingly important (Sarkis 2021). It is important to note that the evolving landscape of humanitarian operations demands a continuous reassessment of supply chain strategies to ensure effective and timely assistance to those in need. By examining the types of disruptions, relevant theories, successful case studies, and emerging trends, researchers and practitioners can contribute to building more resilient and adaptive humanitarian supply chains capable of meeting future challenges. The next section will discuss the type of disruption in supply chain context.

1.2 Type of Disruption in Supply Chain Management

Supply chain disruptions can take many forms, each with unique characteristics and impacts on operations. This review examines several key types of disruptions that affect food supply chains. Cyber attacks have emerged as a significant threat to supply chain operations in recent years. As supply chains become increasingly digitalized and interconnected, they are more vulnerable to malicious actors exploiting vulnerabilities in information systems. Cyber attacks can compromise data integrity, disrupt operations, and lead to financial losses. For example, ransomware attacks can paralyze entire supply chain networks by encrypting critical data and demanding payment for its release. The impacts of such attacks can be far-reaching, affecting not only the targeted organization but also its partners and beneficiaries throughout the supply chain. Kazancoglu et al. (2022) highlight the importance of cybersecurity measures in humanitarian supply chains, especially as digital technologies are increasingly adopted to enhance efficiency and transparency. Implementing robust cybersecurity protocols and contingency plans is crucial for mitigating these risks.

Man-made disasters, such as industrial accidents, terrorist attacks, or armed conflicts, can cause sudden and severe disruptions to supply chains. These events often result in damage to critical infrastructure, loss of life, and displacement of populations, all of which pose significant challenges to humanitarian operations. For instance, the 2011 Fukushima nuclear disaster in Japan led to widespread supply chain disruptions across multiple industries, affecting global supply networks (Koshimura and Shuto 2015). In humanitarian contexts, man-made disasters can complicate aid delivery by creating access restrictions, safety concerns for aid workers, and increased demand for specialized resources. Weldegiargis et al. (2023) examined the impact of armed conflict on household food insecurity in Tigray, Ethiopia, demonstrating how such events can severely disrupt food supply chains and threaten food security for vulnerable populations.

Climate disruptions are becoming more frequent and severe, posing long-term challenges to supply chain resilience. These disruptions can manifest as acute events, such as hurricanes or floods, or chronic stresses, like sea-level rise or changing precipitation patterns. Climate change impacts can affect all aspects of the supply chain, from sourcing and production to transportation and distribution. Hendry et al. (2019) emphasize the need for supply chains to adapt to climate-related risks, suggesting that localization strategies may enhance resilience in the face of climate uncertainties. Malik et al. (2022) estimated the spillover effects regarding social and health impacts resulting from extreme weather events in food supply chains, highlighting the far-reaching consequences of climate disruptions.

Geopolitical disruptions, including trade disputes, sanctions, and political instability, can significantly impact the flow of goods and services across borders. These disruptions often lead to changes in regulations, tariffs, and trade agreements, which can increase costs and create uncertainty in supply chains. Davarzani et al. (2015) examined the impact of economic sanctions on automotive supply chains, highlighting how geopolitical factors can force organizations to reconfigure their supply networks. In humanitarian contexts, geopolitical tensions can hinder access to certain regions and complicate cross-border operations, affecting the delivery of aid to vulnerable populations. Li et al. (2023) proposed a mathematical model to optimize the transport system in a green containerized grain supply chain in Ukraine, considering disruption scenarios related to the ongoing conflict.

Natural disasters, such as earthquakes, tsunamis, and volcanic eruptions, can cause sudden and severe disruptions to supply chains. These events often result in widespread damage to infrastructure, communication networks, and transportation systems, making it challenging to deliver aid and resources to affected areas. Day (2014) explored the concept of Complex Adaptive Supply Networks (CASN) in the context of disaster relief, emphasizing the need for flexibility and adaptability in responding to natural disasters. The unpredictable nature of these events requires humanitarian organizations to maintain robust preparedness measures and rapid response capabilities. Hasnain et al. (2023) explored the behavior of the Socio-ecological System of the Food Bank of Central and Eastern North Carolina during Hurricane Florence, providing insights into how food banks adapt to natural disaster disruptions.

Financial crises can have far-reaching effects on supply chains, disrupting cash flows, altering exchange rates, and potentially leading to the collapse of key suppliers or partners. Economic downturns can reduce funding for humanitarian operations, forcing organizations to do more with fewer resources. The 2008 global financial crisis, for instance, had significant implications for humanitarian supply chains, affecting donor contributions and the ability of organizations to maintain their operations (Kovács and Spens 2007). Financial disruptions underscore the importance of diversifying funding sources and maintaining financial reserves to ensure the continuity of humanitarian operations during economic uncertainties. Gholami-Zanjani et al. (2021) developed a generic mathematical model with mixed integers to integrate inventory placement, allocation, and supply decisions, seeking to maximize expected total utility and minimize total operating costs in the face of financial uncertainties.

Supply disruptions can occur due to various factors, including production issues, quality problems, or capacity constraints. These disruptions can lead to shortages of critical items, increased lead times, and higher costs. During the COVID-19 pandemic, global supply chains experienced unprecedented disruptions, with shortages of essential items like personal protective equipment (PPE) affecting humanitarian operations worldwide (Travers and Lowe 2021). McDougall and Davis (2024) highlight how the pandemic led to a reevaluation of supply chain strategies, with many organizations turning to local sourcing and production to mitigate disruptions. Below Fig. 1.1 shows the various types of supply chain disruptions discussed in the literature review.

In conclusion, understanding the various types of supply chain disruptions is crucial for developing comprehensive risk management strategies in humanitarian operations. By recognizing the unique characteristics and potential impacts of each type of disruption, organizations can better prepare for and respond to challenges, ultimately improving their ability to deliver critical aid and resources to those in need. As the landscape of humanitarian operations continues to evolve, ongoing research and innovation will be crucial in building more resilient and adaptive supply chains capable of withstanding future disruptions. Additionally, understanding

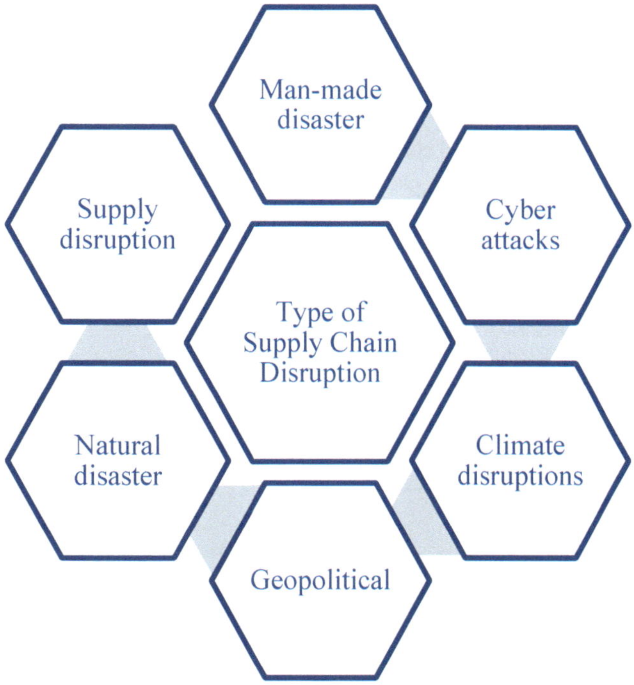

Fig. 1.1 Type of supply chain disruption. *Source* The authors

supply chain disruptions significantly impact humanitarian activities, necessitating innovative interventions. For instance, these disruptions challenge the delivery of critical aid, forcing organizations to adapt rapidly. Therefore, in response, humanitarian actors have developed to innovative solutions such as localizing supply networks to reduce dependency on global chains (McDougall and Davis 2024); implementing digital technologies for improved tracking and coordination; developing resilient funding mechanisms to withstand financial shocks (creating flexible distribution systems to navigate geopolitical barriers and utilizing predictive analytics for better disaster preparedness (Day 2014).

These interventions enhance the agility and resilience of humanitarian supply chains, enabling more effective responses to diverse disruptions. At the same time, recent study by Rahman et al (2022) shows the trend of technology application in examining humanitarian activities in the supply chain before and after crisis era of COVID-19. The next section will discuss on past studies and relevant technology application used in humanitarian efforts.

1.3 Past Studies of Technology Application in Supply Chain Disruption in Humanitarian Operations

Past studies of technology application in supply chain disruption management for humanitarian operations have revealed innovative approaches to enhance resilience and responsiveness. Researchers have examined various technologies, including blockchain for improved traceability (Yadav et al., 2021), Internet of Things (IoT) for real-time monitoring (Singh et al. 2021), and artificial intelligence for predictive analytics (Tsiamas 2021). These studies highlight how digital technologies can strengthen humanitarian supply chains by improving visibility, coordination, and decision-making processes. As disruptions become more frequent and complex, understanding the role of technology in mitigating their impacts is crucial for developing more effective and adaptive humanitarian supply chain strategies.

Figure 1.2 above shows the bibliometric analysis of 243 papers from the Scopus

Basic Info.	
Start Year	1998
End Year	2024
Total Publications	243
Number of Contributing Auth‹	616
Number of Cited Papers	146
Total Citations	3,192
Citation per Paper	13.14
Citation per Cited Paper	21.86
Citation per Author	5.18
Citation sum within h-Core	3,036
Citable Year	27
h-index	26
g-index	53
Publication Years	1998 - 2024
Citation Years	26
Citation per Year	122.77
Author per Paper	2.53
m-index	0.963

Main Information	Data
Publication Years	1998 - 2024
Total Publications	243
Citable Year	27
Number of Contributing Authors	616
Number of Cited Papers	146
Total Citations	3,192
Citation per Paper	13.14
Citation per Cited Paper	21.86
Citation per Year	122.77
Citation per Author	5.18
Author per Paper	2.53
Citation sum within h-Core	3,036
h-index	26
g-index	53
m-index	0.963

Source: Generated by the author(s) using biblioMagika® (Ahmi, 2024)

Document Type	TP	%
Article	78	32.10%
Conference Paper	66	27.16%
Conference Review	40	16.46%
Book Chapter	26	10.70%
Review	12	4.94%
Note	9	3.70%
Editorial	6	2.47%
Book	4	1.65%
Erratum	2	0.82%

Source Type	TP	%	
Journal	108	44.44%	
Conference Proceeding	99	40.74%	
Book	25	10.29%	
Book Series	8	3.29%	
Trade Journal	2	0.82%	
Undefined	1	0.41%	0.00%

Language	TP	%
English	233	95.88%
Russian	7	2.88%
Spanish	3	1.23%
Norwegian	1	0.41%

Fig. 1.2 Basic information on the 243 articles discussing technologies for humanitarian supply chains (scopus data, n = 243). *Source* The authors

database, spanning 1998–2024, offers valuable insights into the evolving landscape of technology applications in humanitarian supply chains. This dataset reveals a sustained scholarly interest in the field over nearly three decades, with a notable acceleration in recent years. The publication timeline indicates a gradual increase in research output, with the field gaining significant momentum in the latter half of the study period. This trend aligns with the broader digital transformation across industries and the growing recognition of technology's potential to enhance humanitarian operations. The average of 9.23 papers per year suggests a consistent, albeit moderate, research focus. Collaboration appears to be a hallmark of this research domain, as evidenced by the substantial number of contributing authors (616) relative to the total publications. This collaborative nature likely stems from the interdisciplinary character of humanitarian supply chain studies, which often integrate expertise from logistics, information systems, and disaster management. The impact of this body of work is reflected in the citation metrics. With 3152 total citations and an average of 13.14 citations per paper, these studies demonstrate considerable influence within the academic community. The h-index of 27 further underscores the presence of highly cited works that have shaped discourse in the field.

Examining document types reveals a clear preference for journal articles (70.78%) as the primary medium for disseminating research findings by the authors. This preference for peer-reviewed journals aligns with the rigorous standards expected in academic publishing. Conference papers (27.16%) represent the second most common document type, highlighting the role of academic conferences in fostering knowledge exchange and collaboration in this domain. The language distribution overwhelmingly favors English (95.88%), which is unsurprising given its status as the lingua franca of international scientific communication. However, the presence of publications in Russian, Spanish, and Portuguese, albeit in small numbers, indicates some degree of global engagement with the topic beyond Anglophone academia. Source types predominantly consist of journals (60.91%) and conference proceedings (32.51%), reinforcing the observation that researchers in this field prioritize venues that offer peer review and opportunities for scholarly dialogue. The minimal representation of book series and trade publications suggests that the rapid pace of technological advancement may favor more agile publication formats. This bibliometric overview provides a foundation for understanding the scholarly landscape of technology applications in humanitarian supply chains, offering insights into research trends, collaboration patterns, and knowledge dissemination practices in this critical field.

Table 1.1 and Fig. 1.3 below show the top 20 sources of publication on technology in humanitarian supply chain. The analysis of publication sources reveals a diverse academic landscape for research on technology in humanitarian supply chains. Social Sciences dominates with 48.15% of publications, underscoring the field's strong societal focus. Computer Science and Engineering follow closely, comprising 33.33% and 30.86% respectively, reflecting the technical underpinnings of supply chain innovations. Business, Management, and Accounting, along with Energy, each contribute 16.05%, highlighting the operational and resource-centric aspects of the domain. The presence of Decision Sciences (9.88%) and Environmental Science (9.47%) indicates a growing emphasis on data-driven decision-making and sustainability within humanitarian logistics. Interestingly, Medicine (8.64%) and Arts and Humanities (8.23%) also feature prominently, suggesting an interdisciplinary approach that considers both health outcomes and cultural contexts. The inclusion of diverse fields such as Agricultural Sciences, Physics, and Economics demonstrates the multifaceted nature of humanitarian supply chain challenges, requiring insights from various disciplines to develop comprehensive technological solutions. Additionally, Table 1.2 explains the top 20 highly cited articles in the field of technology in humanitarian supply chain context.

Table 1.1 Source of publication on technology application in humanitarian supply chain from scopus database (n = 243)

Subject area	Total publication	Percentage (%)
Social sciences	117	48.15
Computer science	81	33.33
Engineering	75	30.86
Business, management and accounting	39	16.05
Energy	39	16.05
Decision sciences	24	9.88
Environmental science	23	9.47
Medicine	21	8.64
Arts and humanities	20	8.23
Agricultural and biological Sciences	14	5.76
Physics and astronomy	10	4.12
Economics, econometrics and finance	9	3.70
mathematics	9	3.70
Earth and planetary sciences	6	2.47
Health professions	5	2.06
Nursing	4	1.65
Biochemistry, genetics and molecular biology	3	1.23
Chemical engineering	2	0.82
Materials science	2	0.82
immunology and microbiology	1	0.41
Multidisciplinary	1	0.41
Psychology	1	0.41

Source The authors

Fig. 1.3 Top 20 sources of publication on Technology and Humanitarian Supply Chain Study. *Source* The authors

Blockchain Technology and Supply Chan in Humanitarian

Blockchain technology presents a promising solution to address critical challenges in humanitarian supply chain management. This distributed ledger system offers enhanced transparency, traceability, and trust—key attributes often lacking in complex humanitarian operations involving multiple stakeholders (Hunt et al., 2023). By providing an immutable record of transactions and product movements, blockchain can significantly improve the tracking of relief goods from donors to beneficiaries, potentially reducing issues like fraud and inefficiencies in distribution (Izadi et al., 2023). The technology's ability to facilitate smart contracts enables automated and transparent execution of agreements between parties, potentially streamlining processes like fund transfers and resource allocation (Dutta et al., 2020). In disaster response scenarios, where rapid coordination is crucial, blockchain-based systems can enable real-time information sharing and improved collaboration among relief agencies (Hunt et al., 2023). For instance, blockchain implementations have shown promise in enhancing the traceability of medical supplies and facilitating cash-based assistance programs (Dutta et al., 2020). However, the adoption of blockchain in humanitarian contexts faces challenges, including the need for technological infrastructure, data standardization, and organizational readiness (Izadi et al., 2023). As the technology matures, careful consideration must be given to issues of data privacy, scalability, and integration with existing systems. Despite these hurdles, blockchain's potential to address longstanding issues of accountability and efficiency in humanitarian aid delivery makes it a compelling area for further research and pilot implementations (Hunt et al., 2023).

Internet of Things and Supply Chain in Humanitarian

The Internet of Things (IoT) has emerged as a transformative technology with significant potential to enhance humanitarian supply chain operations. By connecting physical objects to digital networks, IoT enables real-time data collection, monitoring, and analysis, which can greatly improve the efficiency and effectiveness

Table 1.2 Top 20 highly cited articles

No	Author/year	Title of article	Source
1	Conference Review (2023)	A review of papers from 2023 IEEE 11th region 10 humanitarian technology conference, IHTC 2023	IEEE Region 10 humanitarian technology conference, IHTC
2	Rejali S; Heiniger Y (2020)	The role of digital technologies in humanitarian law, policy and action: Charting a path forward	International review of the red cross
3	Kostelnick JC (2023)	Humanitarian relief and geospatial technologies	The routledge handbook of geospatial technologies and society
4	Usmani UA; Happonen A; Watada J (2023)	Secure integration of IOT-enabled Sensors and technologies: Engineering Applications for humanitarian impact	HORA 2023—2023 5th international congress on human–computer interaction, optimization and robotic applications, proceedings
5	Wynsberghe AV (2020)	Care ethics, philosophy of technology, and robots in humanitarian action	The oxford handbook of philosophy of technology
6	(2023)	2023 IEEE international humanitarian technology conference, IHTC 2023	2023 IEEE international humanitarian technology conference, IHTC 2023
7	Saad S; Maina J; Perera T; Bahadori R (2022)	Blockchain technology-understanding its application in humanitarian supply chains	Advances in transdisciplinary engineering
8	Marić J; Galera-Zarco C; Opazo-Basáez M (2022)	The emergent role of digital technologies in the context of humanitarian supply chains: A systematic literature review	Annals of operations research
9	Tukhvatulina L (2023)	Fast technologies and slow institutions about the place of social and humanitarian knowledge in understanding of the prospects of AI	Filosofiya. Zhurnal Vysshey Shkoly Ekonomiki
10	Edwards D; Subramanian N; Chaudhuri A; Morlacchi P; Jia F; Zheng X; Chen L (2025)	Use of delivery drones for humanitarian operations: Analysis of adoption barriers among logistics service providers from the technology acceptance model perspective	Annals of operations research
11	(2020)	Proceedings Of B-HTC 2020—1st IEEE Bangalore humanitarian technology conference	Proceedings of B-HTC 2020—1st IEEE Bangalore humanitarian technology conference
12	Bodrunov SD (2022)	Anthology of noonomy: Fourth technological revolution and its economic, social and humanitarian consequences: Technology and socio-economic progress: Traps and opportunities for the future	Anthology of noonomy: Fourth technological revolution and its economic, social and humanitarian consequences: Technology and socio-economic progress: Traps and opportunities for the future

(continued)

Table 1.2 (continued)

No	Author/year	Title of article	Source
13	(2020)	Testimonies: How humanitarian technologies impact the lives of affected populations	International review of the red cross
14	Lavonne Reimer JD. (2022)	Humanitarian technology due diligence, Part II, United Nations SDGs 5 (Gender Equity) And 8 And 10 (Economic Equality)	2022 IEEE global humanitarian technology conference, GHTC 2022
15	Sekara V; Karsai M; Moro E; Kim D; Delamonica E; Cebrian M; Luengo-Oroz M; Jiménez RM; Garcia-Herranz M (2024)	The opportunities, limitations, and challenges in using machine learning technologies for humanitarian work and development	Advances in complex systems
16	Izadi E; Nikbakht M; Feylizadeh MR; Shahin A (2023)	A system dynamics model in the humanitarian supply chain based on blockchain technology	International journal of disaster risk reduction
17	Conference Review (2024)	2024 IEEE 1st Karachi section humanitarian technology conference, Khi-HTC 2024	2024 IEEE 1st Karachi section humanitarian technology conference, Khi-HTC 2024
18	Herath T (2024)	The Use of technology in global health humanitarian settings: How Theory Plays out in practice	Health and technology
19	Usova JV; Dushina TV; Chagilov VR; Semkina EN; Masalov AG (2022)	Humanitarian technologies in the education system of modern Russia	Advances in science, technology and innovation
20	Khan M; Imtiaz S; Parvaiz GS; Hussain A; Bae J (2021)	Integration of Internet-Of-Things with blockchain technology to enhance humanitarian logistics performance	IEEE access

Source The authors

of disaster response and aid distribution efforts. In humanitarian contexts, IoT devices such as sensors, RFID tags, and GPS trackers can be deployed to monitor the location, condition, and movement of relief supplies throughout the supply chain (Dutta et al., 2020). This capability is particularly valuable in tracking perishable goods like food and medicine, ensuring they reach beneficiaries in optimal condition. Moreover, IoT technology can facilitate more accurate demand forecasting and inventory management, helping humanitarian organizations to allocate resources more effectively in rapidly changing crisis environments (Izadi et al., 2023). The integration of IoT with blockchain technology presents further opportunities for enhancing transparency and trust in humanitarian supply chains. By combining the data collection capabilities of IoT with the immutable ledger of blockchain, relief organizations can create a more secure and auditable record of aid distribution, potentially reducing fraud and improving accountability to donors and beneficiaries alike (Hunt et al., 2023; Khan et al, 2021). However, the implementation of IoT in humanitarian settings faces several challenges, including the need for reliable connectivity in disaster-stricken areas, data security concerns, and the cost of deploying and maintaining IoT infrastructure. Despite these obstacles, the potential benefits of IoT in terms of improved operational efficiency, enhanced situational awareness, and increased supply chain visibility make it a promising area for further research and development in the humanitarian sector. As natural disasters and complex emergencies continue to pose significant challenges to global humanitarian efforts, the integration of IoT technologies may prove crucial in building more resilient and responsive supply chains capable of delivering aid more effectively to those in need.

UAV and Drone Technology in Supply Chain Humanitarian

Drone technology has emerged as a pivotal tool in humanitarian supply chains (HSC), offering innovative solutions to many logistical challenges faced in disaster response and relief efforts. Drones, or unmanned aerial vehicles (UAVs), have the capacity to overcome obstacles such as inaccessible terrain, damaged infrastructure, and disrupted communication networks, providing swift delivery of critical supplies like medical aid, food, and water to isolated or disaster-stricken areas. This speed and flexibility in logistics have significantly enhanced the efficiency of relief efforts, particularly in time-sensitive situations.

In terms of operational capabilities, drones excel in transportation and delivery, surveying, and real-time monitoring. Their ability to bypass road networks reduces lead times, while their capacity to carry lightweight, high-value items like medical supplies ensures timely interventions in emergencies. Drones can also survey vast areas affected by disasters, capturing high-resolution imagery for damage assessment and monitoring, which is crucial for situational awareness and the effective allocation of resources (Rejeb et al. 2021; Zhang et al. 2023; and in et al. 2023).

Another critical aspect of drone deployment in HSC is their communication and integration capabilities. Drones can serve as aerial communication hubs, facilitating the establishment of ad hoc networks when terrestrial communication systems are down. This enhances coordination between rescue teams, non-governmental organizations, and affected communities, improving overall disaster response efforts (Rejeb et al. 2021). Despite their advantages, the integration of drones into HSC faces challenges such as regulatory issues, privacy concerns, and technological barriers. However, as advancements in drone technology continue, their role in humanitarian logistics is expected to expand, providing cost-efficient, scalable, and sustainable solutions for future disaster response efforts.

AI and Machine Learning Technology in Supply Chain Humanitarian

Artificial Intelligence (AI) and Machine Learning (ML) have proven instrumental in enhancing the effectiveness of humanitarian supply chains (HSC), especially in disaster relief operations. AI-driven systems improve decision-making and resource allocation, addressing the complexities of demand unpredictability, location-specific challenges, and rapid response needs. Through predictive analytics, machine learning can forecast disaster impacts, enabling agencies to optimize inventory pre-positioning and supplier selection, ensuring critical supplies are available when needed. This results in significant cost savings and risk mitigation (Dubey et al. 2022).

One key application of AI in HSC is in the area of supplier selection. Traditionally, selecting suppliers in a humanitarian context involves balancing cost, delivery time, and supply resilience. AI systems can process large datasets to evaluate these factors, helping relief organizations make informed, data-driven decisions. Machine learning algorithms, such as those integrating stratified random sampling and K-means clustering, enhance computational efficiency in selecting optimal suppliers based on fluctuating demand and supply conditions. For example, Hu et al. (2024) proposed a multi-objective stochastic programming model that uses machine learning-based sample average approximation to optimize supplier selection while reducing procurement and transportation costs.

AI also plays a role in real-time decision-making during relief efforts. Drones equipped with AI can monitor disaster zones, providing live data for logistics planning and supply chain optimization. This technology reduces the time needed for damage assessment, thereby speeding up the delivery of essential goods to affected areas. Moreover, AI systems help manage complex supply chain networks by predicting potential disruptions and offering alternative routing and supply strategies. This fosters resilience in humanitarian operations, ensuring aid reaches its destination despite challenges. In conclusion, AI and machine learning are indispensable in modernizing and streamlining humanitarian supply chains, improving both preparedness and response in disaster scenarios.

Future Research and Conclusion

Future research in humanitarian logistics and supply chain management should prioritize several key areas to address emerging challenges and capitalize on technological advancements. A critical focus is the need for more empirical studies examining the long-term impacts of innovative technologies, such as blockchain, Internet of Things (IoT), and artificial intelligence (AI), on humanitarian operations. These technologies offer transformative potential but require rigorous evaluation to understand their full implications and optimize their implementation in complex humanitarian contexts. Specifically, these studies should assess not only the operational benefits but also the social and ethical implications of technology adoption in crisis or disaster environments. Additionally, researchers should explore the integration of multiple technologies to create more robust and adaptive humanitarian supply chains. For instance, combining blockchain with IoT and AI could potentially revolutionize end-to-end visibility, predictive analytics, and decision-making processes in complex humanitarian contexts. Another critical area for future scholars is the development of localized and sustainable supply chain models that can withstand global disruptions while empowering local communities. This includes investigating strategies for building resilient local production capacities, fostering cross-sector partnerships, and implementing circular economy principles in humanitarian supply chains. Furthermore, as climate-related disasters become increasingly frequent and severe, there is an urgent need for research on climate-resilient humanitarian logistics strategies, including adaptive infrastructure design and innovative last-mile delivery solutions for extreme environments. These frontiers of research will significantly enhance the current body of literature on humanitarian supply chains, aligning with the growing imperative for sustainable and adaptable humanitarian responses to emerging challenges.

In conclusion, this chapter has highlighted the critical role of technology in enhancing the resilience and effectiveness of humanitarian supply chains. From blockchain's potential to improve transparency and traceability to the use of AI and machine learning in optimizing resource allocation and decision-making, technological innovations are reshaping the landscape of humanitarian operations. However, the successful implementation of these technologies requires careful consideration of contextual factors, including infrastructure limitations, data privacy concerns, and the need for capacity building among humanitarian actors. As the field continues to evolve, it is crucial for researchers and practitioners to collaborate closely, ensuring that technological advancements are aligned with the core humanitarian principles of humanity, neutrality, impartiality, and independence. By bridging the gap between theoretical research and practical application, the humanitarian sector can develop more agile, efficient, and impactful supply chains capable of meeting the complex challenges of future crises.

References

Cardoso SR, Barbosa-Póvoa AP, Relvas S, Novais AQ (2016). Evaluating supply chain resilience under different types of disruption. In: computational management science: State of the Art 2014 pp 123–129. Springer International Publishing

Conference review - Lozano, Carlos, and Andres Navarro. Welcome to the Proceedings of the IEEE International Humanitarian Technology Conference (IHTC) 2023!. (2023).

Davarzani H, Zanjirani Farahani R, Rahmandad H (2015). Understanding econo-political risks: Impact of sanctions on an automotive supply chain. Int. J. Oper. Prod. Manag. 35(11):1567–1591

Day JM (2014). Fostering emergent resilience: The complex adaptive supply network of disaster relief. International J. Prod. Res. 52(7):1970–1988

Dubey R, Bryde DJ, Dwivedi YK, Graham G, Foropon C (2022). Impact of artificial intelligence-driven big data analytics culture on agility and resilience in humanitarian supply chain: A practice-based view. Int. J. Prod. Econ. 250:108618

Dutta P, Choi TM, Somani S, Butala R (2020). Blockchain technology in supply chain operations: Applications, challenges and research opportunities. Transportation research part e: Logistics and transportation review, 142:102067

Fujita M, Hamaguchi N (2016). Supply chain internationalization in East Asia: Inclusiveness and risks. Papers in regional science, 95(1):81–100

Gholami-Zanjani SM, Jabalameli MS, Klibi W, Pishvaee MS (2021). A robust location-inventory model for food supply chains operating under disruptions with ripple effects. Int. J. Prod. Res. 59(1):301–324

Hasnain T, Walton TN, Odubela K, McConnell S, Davis L, Ivy J, Ivy S Jiang, Coan D, Islam MH, Mpere E (2023). Resiliency within the socio-ecological system of a large food bank network: preparing, mitigating, responding, and recovering from hurricane florence. Int. J. Disaster Risk Reduct. 88:103580

Hendry LC, Stevenson M, MacBryde J, Ball P, Sayed M, Liu L (2019). Local food supply chain resilience to constitutional change: The Brexit effect. Int. J. Oper. Prod. Manag. 39(3):429–453

Herlin H, Pazirandeh A (2015). Avoiding the pitfalls of cooperative purchasing through control and coordination: Insights from a humanitarian context. Int. J. Procure. Manag. 8(3):303–325

Hu S, Dong ZS, Dai R (2024). A machine learning based sample average approximation for supplier selection with option contract in humanitarian relief. Transportation research part E: logistics and transportation review, 18:103531

Hunt K, Narayanan A, Zhuang J (2022). Blockchain in humanitarian operations management: A review of research and practice. Socio-economic planning sciences, 80:101175

Hunt M, Pringle J, Christen M, Eckenwiler L, Schwartz L, & Davé A (2016). Ethics of emergent information and communication technology applications in humanitarian medical assistance. Int. Health. 8(4):239–245.

IHTC 2023 - Lozano, Carlos, and Andres Navarro. Welcome to the Proceedings of the IEEE International Humanitarian Technology Conference (IHTC) 2023!. (2023).

Izadi E, Nikbakht M, Feylizadeh MR, Shahin A (2023). A system dynamics model in the humanitarian supply chain based on blockchain technology. Int. J. Disaster Risk Reduct. 96:103977

Jia F, Zheng X, Chen L (2025). Disaster management in supply chains: a systematic literature review with bibliometric and content analysis. Int. J. Logist. Res. Appl. 1–30.

Kapoor K, Bigdeli AZ, Dwivedi YK, Raman R (2021). How is COVID-19 altering the manufacturing landscape? A literature review of imminent challenges and management interventions. Annals of operations research. https://doi.org/10.1007/s10479-021-04397-2

Kazancoglu I, Ozbiltekin-Pala M, Mangla SK, Kazancoglu Y, Jabeen F (2022). Role of flexibility, agility and responsiveness for sustainable supply chain resilience during COVID-19. J. Clean. Prod. 362:132431.

Khan M, Imtiaz S, Parvaiz GS, Hussain A, Bae J (2021). Integration of internet-of-things with blockchain

technology to enhance humanitarian logistics performance. IEEE Access, 9:25422–25436
Kovács G, Falagara Sigala I (2021). Lessons learned from humanitarian logistics to manage supply chain disruptions. J. Supply Chain Manag. 57(1):41–49
Koshimura S, Shuto N (2015). Response to the 2011 great East Japan earthquake and tsunami disaster. Philosophical Transactions of the royal society A: Mathematical, physical and engineering sciences, 373(2053):20140373
Lorenc A, Kuźnar M (2021). The most common type of disruption in the supply chain-evaluation based on the method using artificial neural networks. Int. J. Shipp. Transp. Logist. 13(1–2):1–24
Malik A, Khan KIA, Qayyum S, Ullah F, Maqsoom A (2022). Resilient capabilities to tackle supply chain risks: managing integration complexities in construction projects. Buildings. 12(9):1322.
Marić J, Galera-Zarco C, Opazo-Basáez M (2022). The emergent role of digital technologies in the context of humanitarian supply chains: A systematic literature review. Ann. Oper. Res. 319(1):1003–1044.
McDougall N, Davis A (2024). The local supply chain during disruption: Establishing resilient networks for the future. J. Clean. Prod. 462:142743
Rahman NAA, Ahmi A, Jraisat L, Upadhyay A. (2022). Examining the trend of humanitarian supply chain studies: pre, during and post COVID-19 pandemic. J. Hum.Itarian Logist. Supply Chain. Manag. 12(4):594–617
Rejali S, Heiniger Y (2020). The role of digital technologies in humanitarian law, policy and action: charting a path forward. International review of the Red Cross, 102(913):1–22.
Rejeb A, Rejeb K, Simske S, Treiblmaier H (2021). Humanitarian drones: A review and research agenda. Internet of Things, 16:100434
Saad S, Maina J, Perera T, Bahadori R (2022). Blockchain technology–understanding its application in humanitarian supply chains. In Advances in Manufacturing Technology XXXV (pp. 385–391). IOS Press.
Saïah F, Vega D, deVries H, Kembro J (2023). Process modularity, supply chain responsiveness, and moderators: The Médecins Sans Frontières response to the Covid-19 pandemic. Prod. Oper. Manag. 32(5):1490–1511
Sarkis J (2021). Supply chain sustainability: Learning from the COVID-19 pandemic. Int. J. Oper. & Prod. Manag. 41(1):63–73
Singh S, Kumar R, Panchal R, Tiwari MK (2021). Impact of COVID-19 on logistics systems and disruptions in food supply chain. Int. J. Prod. Res. 59(7):1993–2008
Travers CL (2024). Leveraging local capital in humanitarian supply networks during pandemic response. Int. J. Disaster Risk Reduct. 106:104421
Travers L, Lowe A (2021). Humanitarian supply chains during COVID-19: Systems failures, recovery and emerging alternatives. Engineering X (Report)
Tsiamas K (2021). A simulation-based framework to improve the resilience of the food supply chains (Doctoral dissertation, Loughborough University)
Usmani UA, Happonen A, Watada J (2023, June). Secure integration of IoT-enabled sensors and technologies: Engineering applications for humanitarian impact. In 2023 5th international congress on human-computer interaction, optimization and robotic applications (HORA) (pp. 1–10). IEEE.
Van Wassenhove LN (2006). Humanitarian aid logistics: Supply chain management in high gear. J. Oper. Res. Soc. 57(5):475–489
Vega D, Arvidsson A, Saïah F (2022). Resilient supply management systems in times of crisis. Int. J. Oper. Prod. Manag.
Yadav S, Singh SP (2021). An integrated fuzzy-ANP and fuzzy-ISM approach using blockchain for sustainable supply chain. J. Enterp. Inf. Manag. 34(1):54–78
Yin Y, Yang Y, Yu Y, Wang D, Cheng TCE (2023). Robust vehicle routing with drones under uncertain demands and truck travel times in humanitarian logistics. Transportation research part B: methodological. 174:102781
Zhang G, Jia N, Zhu N, Adulyasak Y, Ma S (2023). Robust drone selective routing in humanitarian transportation network assessment. European J. Oper. Res. Enterp. Inf. Manag. 305(1):400–428

Dr. Nor Aida Abdul Rahman currently serves as an associate professor at Universiti Kuala Lumpur, Malaysian Institute of Aviation Technology (UniKL MIAT) in Subang, Selangor, Malaysia. Her work has been featured in prestigious journals such as Industrial Marketing Management, Journal of Islamic Marketing, Journal of Humanitarian Logistics and Supply Chain Management, and Journal of Quality and Reliability Management, among others. In addition to her role at UniKL MIAT, she holds visiting appointments as a professor at ITL Trisakti in Jakarta, Indonesia, and as a visiting fellow at the UKM-MPOB Endowment Chair at the National University of Malaysia. Dr. Nor Aida has a distinguished record of scholarship, having authored or co-authored over 100 articles, book chapters, and research books. Her expertise also extends to the publishing realm, where she has contributed to the editorial teams of more than 10 books with Routledge and Springer. She earned her Ph.D. in Management (Supply Chain Management) from Brunel University, London, after completing her Master's and Bachelor's degrees in Business and Administration from the National University of Malaysia. Her dedication to the field is further demonstrated by her roles as an MQA panel auditor, a JAKIM halal-certified trainer, an HRDF-certified trainer, and as a Board Member of Education at the Chartered Institute of Logistics and Transport Malaysia (CILTM).

Dr. Kamran Mahroof is an associate professor of Supply Chain Analytics and program leader for the MSc in the Applied Artificial Intelligence and Data Analytics program, at the University of Bradford, School of Management. He has extensive industry experience having worked in various role for leading organizations, most recently for Morrisons Plc, where he led various productivity initiatives in Logistics. Kamran continues to hold a strong connection in several industries through his ongoing research. His industrial experience has helped shape his research interests, whereby he is actively researching in the areas of Supply Chain Management, Circular Economy, Artificial Intelligence (AI), and the Societal Impact of Emerging Technologies. He regularly publishes in high quality, high impact peer-reviewed journals, such as International Journal of Production Economics and Journal of Business Research and Information Systems Frontiers. Kamran makes regular appearances on national television, radio and in other media and has written several articles for the Press, along with other internationally renowned publications, such as for the World Economic Forum. He is also a Fellow of the Higher Education (FHEA).

Dr. Juliater Simarmata, SE, MM is a lecturer and researcher at the Trisakti Institute of Transportation and Logistics, Jakarta—Indonesia. He has several articles and books on artificial intelligence marketing, destination management organization, and many more published in national and international journals. His research interests include airline marketing, marketing strategy, and consumer behavior. Apart from being a lecturer, he is active as a training consultant and speaker in the fields of Sales/Marketing, Business Management, and Entrepreneurship in various companies in Indonesia.

2 Optimising Humanitarian Logistics: A Lesson Learned for Rapid Response to Supply Chain Disruptions in Crisis Situations

Haliza Mohd Zahari, Ruzaidin Mohammed Zain, Najuwa Mohd Nasir, and Nur Hidayah Azhar

Abstract

It is impossible to prevent disasters from happening. Enhancing coordination among responsible organisations is crucial to effectively carry out logistics operations aimed at minimising the suffering of the impacted population. Directive No. 20 provides a clear and detailed explanation of particular responsibilities and duties assigned to each organisation involved. Nevertheless, as the frequency of disasters increases, there has been an improvement in the cooperation across organisations, despite the lack of understanding. Hence, this study was conducted to examine the effective strategies employed and the valuable insights gained from two cases of flooding that took place in urban settings, specifically Kuala Lumpur, and in rural areas in Baling, Kedah. A total of 16 government officials and affected residents were interviewed. Thematic analysis was utilised to collect precise information pertaining to the logistics operations that took place in both locations. Both case studies have utilised a narrative structure. This study identified information and communication, collaboration and building confidence as crucial components for effectively managing logistics operations.

Keywords

Logistics operation · Flood · Lesson learned · Coordination

2.1 Introduction

Following natural disasters, the effectiveness and swiftness of humanitarian logistics can determine between life and death. Optimising humanitarian logistics entails creating a thorough and robust logistics operation that guarantees a prompt and efficient reaction to disturbances in the supply chain (Chong et al. 2019). To ensure the efficient delivery of necessary goods to affected areas, the logistics network needs to incorporate comprehensive planning, real-time data analysis, flexible supply chain management and effective coordination

H. Mohd Zahari (✉) · R. Mohammed Zain
Strategic and Security Studies Institute, HADR Research Center, National Defence University of Malaysia, Kuala Lumpur, Malaysia
e-mail: haliza.mz@upnm.edu.my

N. Mohd Nasir
Faculty of Business and Communication, Universiti Malaysia Perlis, Arau, Malaysia
e-mail: najuwa@unimap.edu.my

N. Hidayah Azhar
National Defence University of Malaysia, Kuala Lumpur, Malaysia

N. A. A. Rahman et al. (eds.), *Supply Chain in Humanitarian Operations*, Sustainable Development Goals Series,
https://doi.org/10.1007/978-981-96-4426-1_2

among all parties involved (Warnier et al., 2020). By prioritising these factors, organisations can enhance their ability to respond to disasters, minimise delays and ensure that aid reaches those in need as quickly and efficiently as possible.

An essential element of a successful humanitarian logistics operation is advanced planning which involves anticipating potential challenges and creating detailed contingency plans (Chingono and Mbohwa 2023). The plans should provide a detailed explanation of the necessary actions to be followed in different disaster situations, encompassing the logistical aspects of procurement, transporting and distributing relief (Sudan et al. 2023). Proactive planning helps mitigate the impact of disruptions by ensuring that resources and personnel are ready and strategically positioned to respond quickly.

Real-time data analytics are essential for optimising logistics by providing current information on the status of operations and the changing needs of affected communities (Kondraganti et al. 2024). Organisations can make informed decisions, adapt to changing conditions and identify potential bottlenecks in the supply chain by implementing systems that can capture and analyse data in real time (Kondraganti et al. 2024). For example, data on road conditions, weather patterns and supply levels can be used to adjust routes and inventory to ensure the timely delivery of aid. By leveraging data analytics, humanitarian organisations can improve the situational awareness and operational efficiency.

In order to address the unpredictable nature of disaster, agile supply chain management is necessary (Upadhyay et al. 2022). Flexibility in logistics operations enables organisations to promptly adjust to changes, including fluctuations in supply availability, disruptions in transportation networks and changing demand levels. In order to achieve this agility, it is necessary to establish relationships with numerous suppliers, maintain flexible inventory systems and have alternative transportation options (Kamalahmadi et al. 2022). Organisations can more effectively address unforeseen challenges and guarantee that aid arrives at its intended destination without significant delays by establishing a dynamic and adaptable supply chain without significant delays.

Collaboration among various stakeholders is fundamental to a successful humanitarian logistics response (Malik, 2022). Effective coordination between government agencies enhances the efficiency of logistics operations and ensures that resources are used effectively. Joint efforts in planning, resource allocation and information sharing can prevent duplication of efforts and optimise the overall response (Shokr et al. 2022). Building strong partnerships and communication channels facilitates a more cohesive and integrated approach to disaster relief, ultimately improving the impact of humanitarian aid. Optimising humanitarian logistics is pivotal for transforming disaster response from a reactive struggle into a well-coordinated, proactive effort that can save lives and alleviate suffering. Therefore, this study aims to explore the issues and challenges faced in humanitarian logistics in two areas which experienced heavy floods and taking the lesson learned for rapid response to supply chain disruptions in future.

2.2 Literature Review

Humanitarian Logistics

Humanitarian logistics is a comprehensive strategy aimed at the efficient and cost-effective management of the flow and storage of goods, materials, and information from their source to their destination (Thomas and Kopczak, 2005). The specific goal of humanitarian logistics is to alleviate the suffering of vulnerable populations. This role involves a broad spectrum of tasks, such as preparation, planning, procurement, transportation, warehousing, monitoring and tracking. Humanitarian logistics as defined by Muhammad Khan (2019), encompasses the strategic coordination and management of the movement of resources to ensure the timely and cost-effective delivery of essential goods and information to those in need. This process aims to protect lives, enhance livelihoods and safeguard infrastructure and assets.

Sreedharan et al. (2020) stated that humanitarian logistics management is crucial for overseeing the logistics process of delivering aid from donors to affected individuals. The primary focus is on reducing suffering and saving lives, rather than prioritising profits, and emphasising the significance of logistics in several crucial tasks that are necessary for a successful humanitarian response. Humanitarian logistics is essential in the management of various types of disasters, including natural events like earthquakes, tsunamis, hurricanes, tornadoes, floods, landslides, droughts and famines, as well as man-made incidents such as chemical leaks, large-scale explosions, terrorist attacks, refugee crises and war situations. These disasters can happen simultaneously or in combination (Masudin et al., 2019). The need for prompt response, scarcity of resources and significant uncertainty make it crucial. Humanitarian logistics is primarily concerned with ensuring efficiency and promptness, acting as a crucial connection between different stages and procedures in crisis management. Furthermore, it supplies crucial information for post-event analysis to assess the effectiveness of disaster response operations.

Gossler et al. (2020) highlight the importance of adopting a systematic approach to humanitarian initiatives. The costs related to relief activities are significantly influenced by logistics expenses, specifically procurement and transportation. This highlights the importance of implementing cost-effective and efficient logistic techniques. Humanitarian logistics involvement is crucial in disaster relief operations to guarantee the efficient and effective delivery of essential resources like healthcare, food, shelter, water, medicine and sanitation. Humanitarian logistics oversees the coordination of various components of the system, including goods, materials, personnel and infrastructure.

The importance of meticulous logistics planning and coordination is emphasised by the frequent occurrence of overlapping relief distribution and underserved communities that independent organisations in the humanitarian sector often meet. This entails the effective allocation of assistance to the final stage, thorough evaluation of needs and management of transportation, ensuring that it reaches the most susceptible individuals (Schiffling et al. 2020; Yáñez-Sandivari et al. 2021).

The primary goal of humanitarian logistics stays unchanged, regardless of the many operations carried out at different times, to support the survival endeavours of suffering populations. Ernst (2003) proposed the concept of the relief effort which has been expanded upon by more recent research. This dynamic process involves the planning, management and control of the flows of information, services and relief to meet the urgent needs of affected individuals.

HL Roles According to Disaster Phases

Disaster operations are commonly categorised into four distinct phases: mitigation, readiness, response and recovery. The effectiveness of humanitarian logistics is paramount at each phase of emergency aid (National Governors' Association. Center for Policy Research, & United States. Defense Civil Preparedness Agency, 1979). In the mitigation and readiness stages, the swift deployment of disaster management equipment guarantees timely aid to the victims (Ukkusuri et al., 2008). Prior strategic allocation of shelter locations improves the organisation of population evacuations (Widener et al. 2011). During the response phase of a disaster, humanitarian logistics involves several tasks including large-scale evacuation, distribution of relief, relocation of victims, removal of debris, clearance of roads and building of local infrastructure (Özdamar et al. 2015). During the recovery phase, humanitarian logistics prioritises the construction of temporary shelters, restoration of key services, repair of infrastructure and provision of medicines for long-term illness prevention. These endeavours seek to expedite the community's restoration and ability to bounce back after the catastrophic incident. The various stages of a disaster include preparedness, response, recovery and mitigation.

The Preparedness Phase involves HL focusing on strategically placing supplies, creating response plans and organising training and exercises. The objective is to guarantee that logistics systems are prepared to be promptly triggered in the event of a crisis. Strategically placing supplies in advance can greatly decrease reaction times and enhance the efficiency of relief operations (Behl and Dutta, 2019). Simulation exercises and training programmes improve the preparedness of HL teams, ensuring they are adequately equipped to handle real-life crisis events (Halizahari et al. 2023; Negi, 2022). Moreover, the establishment of comprehensive reaction plans that incorporate alternative measures guarantees a prompt and well-coordinated initiation of high-level actions (Baharmand et al. 2020).

The Response Phase encompasses the prompt measures made to meet the requirements of individuals impacted by the catastrophe. During this phase, HL engages in actions such as the conveyance of goods to locations affected by disasters, the allocation of assistance to recipients and the facilitation of logistical operations in collaboration with different agencies and organisations. Swift and effective logistics response can greatly diminish the number of deaths and illnesses by guaranteeing the prompt delivery of vital resources to impacted communities (Chiappetta Jabbour et al. 2019). Conducting a prompt needs assessment is essential for identifying the most pressing demands and allocating resources accordingly (Yuste et al. 2019). Incorporating contemporary technology such as real-time tracking systems, further improves the effectiveness of supply chain activities throughout the reaction phase (Nasir and Mohd Zahari, 2023).

During the recovery phase, HL prioritises the reconstruction of infrastructure, replenishment of supplies and the shift from emergency response to long-term development. This stage entails assessing the efficiency of the logistics response, identifying valuable insights and adopting enhancements to enhance readiness for future calamities (Behl and Dutta, 2019). Reconstructing infrastructure is crucial for returning to normalcy and guaranteeing the enduring viability of impacted communities. Replenishing resources guarantee that logistical systems are ready for upcoming catastrophes, hence reducing response time delays (Stumpf et al. 2023). Conducting thorough post-disaster evaluations enables organisations to evaluate the strengths and limitations and make any necessary improvements (Tay et al. 2022).

Reduction of harm or lessening of negative effects, the mitigation phase involves endeavours that try to avert a disaster, lessen the likelihood of its occurrence or minimise the detrimental impacts. During this step, the logistician must strategise the supply chain and carefully assess the associated risks. The strategy may encompass the diversification of supplier bases, adaptation of inventory management systems, exploration of third-party logistics options and cultivation of robust supplier relationships to strengthen resilience and mitigate disruptions.

Humanitarian Logistics and Relief Coordination

Humanitarian logistics is now widely acknowledged as a crucial and desirable approach to addressing humanitarian crises, with the goal of ensuring the effective distribution of aid and the rehabilitation of populations affected by disasters. Government agencies must be involved and their operations must be coordinated in order to effectively resolve a disaster situation. Agencies coordination in the context of natural disaster response involves the merging of facilities, assets, equipment, personnel and communication to support incident response activities (Njuki Karanja et al. 2015). Efficient coordination necessitates the acquisition of knowledge and resources through the organisation of many sectors. This allows for the achievement of outcomes that cannot be accomplished by a single-sector organisation. Coordination can take advantage of the considerable unpredictability associated with factors such as the number of beneficiaries, supply availability, conditions of supply networks and availability of human resources in disaster response (Njuki Karanja et al. 2015). As stated by Njuki Karanja et al. (2015), when conditions are uncoordinated, it leads to duplication of efforts, inefficient use or wastage of resources, and delays, hindrances or obstructions in relief operations. The absence of coordination leads to inefficient distribution of aid.

Various countries have their own infrastructure and methodology for addressing and managing disasters. Typically, agency coordination occurs at three levels: level I involves local agencies, followed by level II state-level coordination. However, if the disaster becomes severe, the response will be escalated to the national level at level III. At level I, the municipal government takes the lead as the primary responders, overseeing rapid relief operations, evacuations and emergency services. Their responsibilities include evaluating the extent of the damage, offering refuge and collaborating with nearby organisations and communities. The state intervenes in disaster response when the magnitude or intricacy of the situation surpasses the capability of local governments to handle efficiently, as is the case in level II. Level III, often referred to as the national level, is triggered when the magnitude of a disaster exceeds the capacity of state and local authorities, requiring the involvement of federal resources and coordination. This situation might arise when state governments seek aid because they lack sufficient resources, or when certain legal requirements are met (Ahmad et al. 2022). National intervention is also prompted in instances of national security risks, extensive public health crises or catastrophes impacting numerous states. Other countries may participate in disaster response efforts either upon the affected country's request for assistance or when the magnitude of the crisis surpasses the country's ability to handle it independently. The 2004 Indian Ocean tsunami highlighted the necessity for a global response, given the catastrophic magnitude of the tragedy which resulted in extensive loss of life and widespread devastation across numerous nations (Tarantino, 2006). Local and national resources proved inadequate in handling large-scale disasters that were beyond the capabilities of a single nation, leading to the depletion of these resources and necessitating extensive international aid.

2.3 Methods

The research was conducted by utilising qualitative technique. The researchers gathered data by physically visiting both locations and conducting observations, meetings and interviews with a total of 16 government officials and residents who were affected. The objective was to gain a deeper understanding of the situation by specifically focusing on the logistics operations in both sites. The interviews conducted adhere to the established interview protocol (Castillo-Montoya, 2016) however the researcher team does not limit the interviews even if the research participants provide additional information. The interview transcript is transcribed verbatim, and then coding is carried out. Thematic analysis was conducted to determine the logistics operations that were in progress during the occurrence of the disaster. The results from the interview were cross-referenced with the Disaster Report for both areas to ensure objectivity and avoid any potential prejudice. Subsequently, the results obtained from both areas are presented in a coherent and descriptive format, as in Case 1 and Case 2.

2.4 Case Study

Case 1: Providing Logistics Support During Disaster Response in Kuala Lumpur

During the period from late 2021 to early 2022, Malaysia encountered persistent severe rainfall as a result of the monsoon season and elevated tides, resulting in historic flooding in various areas, including Kuala Lumpur. This catastrophe resulted in significant destruction to residential dwellings, automobiles, commercial establishments, industrial and agricultural sectors, as well as public properties and infrastructure. At the height of the floods, more than 70,000 people were forced to leave their homes every day across the entire country, resulting in a total of 54 deaths. A total of 620 people, comprising 170 families, were immediately impacted in Kuala Lumpur and were relocated to temporary shelters. The floods caused an estimated MYR 6.1 billion (USD 1.46 billion) in damages across the country, with Kuala Lumpur alone experiencing losses of over RM32.4 million (Department of Statistics Malaysia, 2023).

In Kuala Lumpur, the response operations were primarily directed towards three crucial aspects of humanitarian logistics: provisioning of food and supplies, establishment of temporary evacuation centres and facilitation of transportation. The Department of Social Welfare (JKM) had a crucial function in allocating vital provisions such as food, water, hygiene kits and other essentials to families who had been displaced. The *Depot Simpanan Keperluan Bencana* (DSKB) at Sungai Buloh functioned as the primary depot for Kuala Lumpur, overseeing the storage and distribution of supplies to evacuation centres and residential areas impacted by disasters. The supplies encompassed essential things such as blankets, clothing, personal hygiene products, dry food items, as well as specialised provisions designed for specific demographics, including medical supplies, baby formula and items for the elderly and the disabled.

Kuala Lumpur City Hall (DBKL) and the Department of Social Welfare (JKM) have jointly created and officially designated 113 temporary evacuation shelters in Kuala Lumpur to provide accommodation for displaced residents. During the height of the floods, five shelters provided housing for a large number of evacuees. To effectively manage these centres, it is necessary to do thorough needs assessment, and establish and follow Standard Operating Procedures (SOPs) for the establishment, operation and closure of the centres. The use of these Standard Operating Procedures (SOPs) resulted in the establishment of uniform processes, leading to increased operational efficiency and a decreased likelihood of errors (Schiffling et al. 2020). The MyInfoBencana digital platform, spearheaded by JKM, enabled instantaneous contact and collaboration among many parties, guaranteeing the exchange of precise information and efficient strategising (Yáñez-Sandivari et al. 2021). During the peak of the floods, the Dewan Kelab Sultan Sulaiman and SJK (T) Saraswathy in Kuala Lumpur accommodated 281 and 564 evacuees, respectively (Department of Statistics Malaysia, 2023).

The District Disaster Operation Control Centre, led by the Mayor of Kuala Lumpur, organized the pooling of resources and transportation from other agencies. The Malaysia Civil Defence Force (APM) served as the essential administrative body for disaster management. An adaptable transportation strategy which integrates various forms of transportation such as land vehicles and boats, was crucial in overcoming infrastructure obstacles and guaranteeing the punctual delivery of goods. The utilisation of real-time tracking and routing technologies enabled the ability to make immediate alterations to routes in response to the prevailing road conditions. This capability played a crucial role in ensuring the efficiency and effectiveness of logistics operations amidst the floods. The Integrated Transport Information System (ITIS), overseen by DBKL, played a crucial role in alleviating traffic congestion and enabling the smooth passage of emergency vehicles through the provision of up-to-date traffic information and route guidance.

Challenges and Action Taken

Throughout the flood response, numerous logistical challenges were faced such as congested routes and road closures caused by submerged highways and landslides. These interruptions impeded the prompt delivery of provisions and the mobility of emergency response teams. In order to surmount these obstacles, the subsequent actions were executed:

Utilisation of ITIS: The DBKL-managed Integrated Transport Information System (ITIS) offers real-time traffic reports, allowing emergency response teams to strategise the most efficient routes and circumvent congested or obstructed zones. This technique was crucial in guaranteeing the efficient circulation of traffic and giving priority to the transit of emergency vehicles.

Flexible transportation plans: The utilisation of several mobility modalities such as land vehicles and boats, enabled relief teams to traverse flooded and unreachable areas. This adaptability guaranteed the delivery of resources to impacted communities, even in the face of challenges related to infrastructure.

Inter-agency coordination: The Disaster Operations Control Centre enabled efficient coordination across diverse entities, facilitating the optimal allocation of resources and staff. This cooperation reduced the repetition of efforts and guaranteed that all locations received the required assistance. The federal government, via the National Disaster Management Agency (NADMA), deployed supplementary resources to bolster state initiatives. Efficient utilisation of resources and prevention of redundant efforts necessitated effective communication and coordination among diverse stakeholders (Rahman et al., 2022).

Community involvement: Local communities and volunteers were crucial in delivering prompt aid and assistance to impacted individuals. Their participation facilitated the connection of disparities in official response initiatives and guaranteed a more extensive dissemination of assistance. The floods that occurred in Kuala Lumpur during the 2021–2022 period highlighted the significance of a well-organised humanitarian logistics framework in handling disruptions in the supply chain during times of crisis. The response activities were much improved by using centralised depots such as DSKB, constructing and effectively managing evacuation centres, and implementing adaptive transportation plans with the assistance of ITIS. These procedures guaranteed the fast delivery of vital assistance to those who required it, thereby minimising the consequences of the catastrophe and expediting the recovery process for affected areas.

Case 2: Humanitarian Logistics Support During Disaster Response in Baling, Kedah

Efficient humanitarian logistics plays a critical role in delivering timely and effective relief to affected communities in the context of disaster response. Kampung Iboi, located in Baling, Kedah, encountered substantial difficulties during severe flood calamities on July 4, 2022. The significance of humanitarian logistics has increased further as a result of the persistent heavy rainfall in low-lying areas. This entails the coordination of resources, the management of supply networks and the guarantee that vital supplies are promptly delivered to those who want them. Comprehending the logistical assistance required during the catastrophe response in Baling highlights the importance of effectively coordinating activities and fostering cooperation across logistics.

The significant rise in flood water levels caused extensive damage to properties and posed significant challenges to local authorities and humanitarian services in Kampung Ibai. The catastrophe resulted in the deaths of three individuals, the complete destruction of seventeen residences, the displacement of 3546 inhabitants and an estimated financial loss amounting to RM25.91 million. The victims were sent to a temporary evacuation facility at Sekolah Menengah Kebangsaan Jerai. A considerable proportion of the impacted population had to be relocated to provisional accommodations as a result of the calamity. Currently, logistics support plays a vital role in disaster response by guaranteeing the prompt and efficient distribution of necessary assistance, effectively managing resources and overcoming infrastructural obstacles to satisfy the urgent requirements of impacted communities.

The relief and rehabilitation agencies offered logistics support which encompassed the transportation and distribution of essential supplies such as food, water, shelter, medical aid and other vital commodities. The relief management organisations encountered difficulties in manoeuvring through the impacted areas, contending with powerful currents at Kupang River and evaluating the magnitude of destruction among the disorder caused by the flash flood. The impaired infrastructure exacerbated the complexity of logistics operations, resulting in delayed response times and adversely affecting the efficacy of relief distribution. In addition, the interruption of crucial supply channels and the necessity to set up interim distribution locations posed logistical challenges, necessitating agencies to quickly adjust their strategies. It was essential to collaborate with local authorities and prioritise the safety of relief workers and affected residents in order to successfully address these problems and efficiently provide assistance. Efficient and timely distribution of supplies during disaster response relies heavily on effective humanitarian logistics. This entails the synchronisation of diverse agencies and their responsibilities in overseeing personnel and resources. Agencies cooperate to ensure that relief is delivered to affected areas in a manner that is both effective and efficient.

Challenges and Action Taken

Exploring the challenges faced by humanitarian logistics agencies greatly enhances our comprehension of the intricacies involved in providing assistance by relief and rehabilitation organisations in times of disaster. Addressing these difficulties is essential for improving the resilience and efficiency of future humanitarian logistics operations. The problems are derived from a scenario that takes place during a flood disaster relief effort, specifically focusing on the victims of the investigated cases.

Limited access to affected areas: The limited access to affected areas is a substantial obstacle for humanitarian logistics responders, as it hinders their capacity to provide vital assistance and relief to disaster-affected communities. Obstacles that can impede the passage of persons, equipment and supplies, including infrastructural damage, road blockages and impassable terrain. In order to surmount these obstacles, emergency responders frequently utilise alternative modes of transportation, create temporary pathways for access and apply technology for remote surveillance and coordination. These measures aid in alleviating access problems, guaranteeing that assistance is promptly and efficiently delivered to people requiring it and properly overseeing the entire logistical support during disaster response.

Disruption in communication between emergency responders and the affected communities: The limited accessibility to impacted areas has posed a significant obstacle to the efficient management of disasters. The lack of effective communication hinders responders from obtaining timely updates and precise information regarding the changing requirements and circumstances of the catastrophe victims. In order to tackle this issue, emergency responders utilise satellite phones, deploy drones for aerial evaluations and establish temporary communication hubs to enhance the dissemination of information. This strategy facilitates the closing of communication gaps, enabling more well-informed decision-making and more efficient distribution of relief, so improving the overall effectiveness of disaster response efforts.

Difficulties in performing the initial assessment: Challenges in conducting the initial evaluation by relevant responders might result in inadequate or incorrect information regarding the needs of affected people, causing substantial gaps in comprehending the logistical requirements. These challenges could include limited access to affected locations, disruptions in communication and insufficient resources for thorough data gathering. Consequently, those who respond to the issue may only acquire an incomplete understanding of it which might impede their capacity to appropriately assess the level of harm, recognise urgent requirements and efficiently strategise and organise the allocation of assistance. The absence of a thorough evaluation hinders the development of specific logistics plans and can result in inefficiencies when providing timely and suitable assistance to individuals requiring it.

Insufficient and inaccurate information about the needs of affected populations: Insufficient and incorrect data regarding the requirements of affected people, caused by challenges in conducting the first assessment by relevant responders, might result in substantial gaps in comprehending the logistical needs. The problems may encompass restricted entry to impacted areas, disruptions in communication and insufficient resources for thorough data gathering. In order to tackle these problems, responders utilise remote sensing technology, mobilise quick assessment teams and cooperate with local community networks to collect and authenticate information. These tactics facilitate the discovery and resolution of gaps in the assessment process, enhancing the precision of damage appraisal, improving the identification of priority needs and optimising the planning and coordination of relief distribution.

2.5 Lesson Learned

Due to the location of the flooded zones—one in an urban area and the other in a rural one—there are variances between the case studies concerns and difficulties. Kuala Lumpur's information channels are better coordinated and more structured in order to handle calamities. When all things have been considered, a number of problems and obstacles can be successfully resolved with thorough knowledge and excellent agency coordination.

Information and Communication

During a crisis, information plays a key role due to the involvement of multiple entities with distinct responsibilities. Comprehensive and precise information facilitates the efficient execution of logistics activities. Collaborators must have unrestricted access to information and should not work in isolation. The presence of well-defined protocols and accessible procedures outlined in Directive No. 20 has facilitated the establishment of a hierarchical system of control at all levels, enabling more consistent implementation of logistics operations. Relevant information has been exchanged and suitable assignments have been distributed among the agencies. In Kuala Lumpur, the presence of comprehensive applications that gather all relevant information has expedited the process of providing aid. In contrast to the situation in Baling, the deficiency in information acquisition has led to an imprecise evaluation of logistical requirements. Nevertheless, the establishment of a disaster operations control centre has mitigated this vulnerability, enabling more effective information changes and facilitating seamless information exchange and communication.

Coordination

According to Directive No. 20, each agency has been assigned to a certain cluster and is responsible for carrying out its designated task in the event of a disaster. Disasters contribute to the improvement of professionalism and the clarification of responsibilities, which in turn facilitates the effective execution of logistic operations. Optimal coordination occurs when a Disaster Operations Control Centre is established, enabling organisations to collaboratively analyse and deploy a collection of resources for the distribution of relief assistance. The implementation of a systematic approach and rigorous monitoring enables all participating agencies to achieve better coordination and facilitate the sharing and receiving of identical information.

Building Trust

Among the qualities that should be emulated is the presence of trust. Informal networks established during training and exercises have fostered a collective understanding among individuals, leading to a recognition of the necessity to evacuate. Consequently, there is a lack of non-compliance when evacuation instructions are issued by the agency. This also indirectly enables logistics operations such as the pre-positioning of supplies, and simplifies the process for agencies to deliver help relief.

2.6 Conclusion

Efficient management of humanitarian logistics (HL) is essential for effectively responding to and recovering from natural catastrophes. This study has examined the complexities of HL by concentrating on supply chain management and coordination among different stakeholders. Through an analysis of the flooding incidents in Kuala Lumpur and the disaster response in Baling, Kedah, there is significance of a strong and effective HL in effectively handling disruptions in the supply chain during times of crisis.

The floods in Kuala Lumpur highlighted the crucial importance of centralised depots such as the *Depot Simpanan Keperluan Bencana* (DSKB) in Sungai Buloh. The Department of Social Welfare (JKM) played a crucial role in minimising the immediate effects on impacted populations by efficiently distributing critical supplies. Kuala Lumpur City Hall (DBKL) and JKM development of 113 temporary evacuation centres highlighted the importance of efficiently maintained shelters in offering a secure refuge for evacuees. Moreover, the Integrated Transport Information System (ITIS) overseen by DBKL played a vital role in alleviating traffic congestion and increasing the mobility of emergency vehicles, consequently improving the overall effectiveness of logistics operations.

The incident in Baling, Kedah, underscored the crucial need of humanitarian logistics in responding to disasters. The extensive destruction of property, loss of lives and relocation of inhabitants emphasised the immediate necessity for efficient coordination and resource allocation by relief organisations. The significance of adaptable and versatile logistics solutions was underscored by obstacles such as manoeuvring through flooded areas, disrupted supply routes and impaired infrastructure. The effective distribution of assistance in Baling exemplified the indispensability of cooperation among municipal officials, humanitarian organisations and community volunteers.

Organisations may successfully and promptly respond to emergencies by combining advanced planning, real-time data analytics, agile supply chain management and collaboration among stakeholders. Proactive planning involves anticipating probable issues and preparing detailed contingency plans to guarantee that resources and staff are prepared for immediate deployment. Real-time data analysis improve the ability of organisations to understand their current situation and operate more efficiently, enabling them to make well-informed decisions and adjust to evolving circumstances. Agile supply chain management, known for its flexibility and agility, guarantees the prompt delivery of aid by effectively managing unforeseen obstacles. Efficient collaboration among governments and local agencies maximises the utilisation of resources, amplifies the overall effectiveness of humanitarian assistance and boosts disaster relief operations through robust partnerships and communication channels.

Ultimately, the optimisation of humanitarian logistics is crucial in order to convert catastrophe response into a well-coordinated and proactive endeavour that has the ability to preserve lives and alleviate suffering. Organisations can greatly improve their ability to provide fast and effective assistance by utilising sophisticated planning, real-time data analysis, flexible supply chain management and engagement with stakeholders. These features not only optimise operations but also ensure that resources are used efficiently, so reducing the effects of crises and providing more effective support to impacted populations. Given the growing complexity of disasters, a strong logistics framework is essential for effectively addressing these difficulties and ensuring that relief efforts are successful.

References

Ahmad S, Bin H, Muzamil S, Zainun NY, Ajman N. N, Sulaiman N, Khahro S H, Rohani M, Mohd S, Mohd B, Ahmad H (2022). Proposed framework for the flood disaster management cycle in Malaysia. *Sustainability (Switzerland)*, 14:1–21

Baharmand H, Comes T, Lauras M (2020). Supporting group decision makers to locate temporary relief distribution centres after sudden-onset disasters: A case study of the 2015 Nepal earthquake. Int. J. Disaster Risk Reduct. 45. https://doi.org/10.1016/j.ijdrr.2019.101455

Behl A, Dutta P (2019). Humanitarian supply chain management: A thematic literature review and future directions of research. *Annals of Operations Research*, 283(1–2):1001–1044. https://doi.org/10.1007/s10479-018-2806-2

Castillo-Montoya M (2016). Preparing for interview research: The interview protocol refinement framework. *Qualitative Report*, 21(5):811–831. https://doi.org/10.46743/2160-3715/2016.2337

Chiappetta Jabbour CJ, Sobreiro VA, Lopes de Sousa Jabbour AB, de Souza Campos LM, Mariano EB, Renwick DWS (2019). An analysis of the literature on humanitarian logistics and supply chain management: Paving the way for future studies. Ann. Oper. Res., 283(1–2), 289–307. https://doi.org/10.1007/s10479-017-2536-x

Chingono TT, Mbohwa C (2023). Humanitarian logistics, identifying and reacting to disasters. In supply networks in developing countries: Sustainable and humanitarian logistics in growing consumer markets. *Emerald Publishing Limited*, 1–17

Chong M, Lazo Lazo, JG, Pereda MC, Machuca De Pina JM (2019). Goal programming optimization model under uncertainty and the critical areas characterization in humanitarian logistics management. J. Hum. Itarian Logist. Supply Chain. Manag.9(1):82–107

Department of statistics Malaysia (2023). Special report on flood impacts in Malaysia 2022. *Department of Statistics Malaysia*, 1–59

Ernst R (2003). The academic side of commercial logistics and the importance of this special issue. Forced Migration Review, 18(1): 5–8.

Gossler T, Wakolbinger T, Burkart C (2020). Outsourcing in humanitarian logistics—status quo and future directions. Int. J. Phys. Distrib. & Logist. Management. https://doi.org/10.1108/ijpdlm-12-2018-0400

Halizahari M, Wong MR, Ahmad NDF, Zain R, Zainol NAH (2023). Scenario planning and simulation in disaster response. Int. J. Adv. Sci. Eng. Inf. Technol. 13(4):1235–1241. https://doi.org/10.18517/ijaseit.13.4.17403

Kamalahmadi M, Shekarian M, Mellat Parast M (2022). The impact of flexibility and redundancy on improving supply chain resilience to disruptions. Int. J. Prod. Res. 60(6):1992–2020

Khan M, Yong LH, Han BJ (2019). A systematic review of performance enhancement of humanitarian logistics through transparency: Current status and perspectives. Int. J. Supply Chain. Manag. 8(2):549.

Khan M, Khan M, Ali A, Khan MI, Ullah I, Iqbal M (2022). Digitalization for fast, fair, and safe humanitarian logistics. Logistics, 6(2), 31.

Kondraganti A, Narayanamurthy G, Sharifi H (2024). A systematic literature review on the use of big data analytics in humanitarian and disaster operations. Ann. Oper. Res. 335(3):1015–1052

Malik H (2022). Malaysia humanitarian logistics: A successful collaboration in humanitarian relief works between NGOs in Malaysia. Malaysian J. Innov. Eng. Appl. Soc. Sci. (MYJIEAS), 2(01): 46–63

Masudin I, Fernanda FW (2019). A review of literature on types, stages of recovery and humanitarian logistics operations in the tsunami and earthquake disaster in Indonesia. In *IOP conference series: Materials science and engineering* 674(1):012043. IOP Publishing.

Mahbubur Rahman M, Sadequr Rahman M, Jerin T (2023) Social vulnerability to earthquake disaster: insights from the people of 48th ward of Dhaka South City, Bangladesh. Environ. Hazards 22(2):116–135.

National Governors' Association. Center for policy research, and United States. Defense civil preparedness agency. (1979). Comprehensive emergency management: A governor's guide. [Department of Defense], Defense Civil Preparedness Agency

Nasir N, Zahari HM, Husain R (2023). A systematic literature review on logistics information needs for sharing in Malaysian disaster management. *Sustainability*. https://doi.org/10.3390/su15054524

Negi S (2022) Humanitarian logistics challenges in disaster relief operations: A humanitarian organisations' perspective. J. Transp. Supply Chain. Management. https://doi.org/10.4102/jtscm.v16i0.691

Njuki Karanja P, Mairura CJ, Kepha O (2015). Determinants of effective logistics coordination among humanitarian organizations in Kenya. A case of humanitarian organization in Nairobi County. Int. J. Sci. Res. Publ. 5(6):327–332. http://www.ijsrp.org/e-journal.html

Özdamar L, Ertem MA (2015) Models, solutions and enabling technologies in humanitarian logistics. European J. Oper. Res. 244(1):55–65

Schiffling S, Hannibal C, Tickle M, Fan Y (2020). The implications of complexity for humanitarian logistics: A complex adaptive systems perspective. Ann. Oper. Res. 319:1379–1410. https://doi.org/10.1007/s10479-020-03658-w

Shokr I, Jolai F, Bozorgi-Amiri A (2022). A collaborative humanitarian relief chain design for disaster response. Comput. Ind. Eng. 172:108643.

Sreedharan V, Kek V, Dhanya M, Anjali S, Arunprasad P (2020). Understanding the role of Logistics in humanitarian operations: Key findings and analysis from literatures. Int. J. Logist. Syst. Management. https://doi.org/10.1504/ijlsm.2020.10030522

Stumpf J, Besiou M, Wakolbinger T (2023). Supply chain preparedness: How operational settings, product and disaster characteristics affect humanitarian responses. Prod. Oper. Manag. 32(8):2491–2509

Sudan T, Taggar R, Jena PK, Sharma D (2023). Supply chain disruption mitigation strategies to advance future research agenda: A systematic literature review. J. Clean. Prod. 425:138643

Tay H, Banomyong R, Varadejsatitwong P, Julagasigorn P (2022). Mitigating risks in the disaster management cycle. Advances in civil engineering. https://doi.org/10.1155/2022/7454760

Tarantino, D. (2006). Asian tsunami relief: Department of Defense public health response: Policy and strategic coordination considerations. Mil. Med. 171(suppl_1):15–18.

Thomas AS, Kopczak LR (2005) From logistics to supply chain management: the path forward in the humanitarian sector. Fritz Institute

Ukkusuri SV, Yushimito WF (2008). Location routing approach for the humanitarian aprepositioning problem. Transp. Res. Rec. 2089(1):18–25

Upadhyay A, Mukhuty S, Kumari S, Garza-Reyes JA, Shukla V (2022). A review of lean and agile management in humanitarian supply chains: Analysing the pre-disaster and post-disaster phases and future directions. Prod. Plan. Control. 33(6–7):641–654

Warnier, M., Alkema, V., Comes, T., & Van de Walle, B. (2020). Humanitarian access, interrupted: Dynamic near real-time network analytics and mapping for reaching communities in disaster-affected countries. Or Spectrum, 42:815–834

Yáñez-Sandivari L, Cortés CE, Rey PA (2021). Humanitarian logistics and emergencies management: New perspectives to a sociotechnical problem and its optimization approach management. Int. J. Disaster Risk Reduct. 52:101952

Yuste P, Campbell JR, Canyon D, Childers M, Ryan B (2019). Synchronized humanitarian, military and commercial logistics: An evolving synergistic partnership. Safety. https://doi.org/10.3390/safety5040067

Widener MJ, Horner MW (2011). A hierarchical approach to modeling hurricane disaster relief goods distribution. J. Transp. Geogr. 19(4):821–828

Dr. Haliza Mohd Zahari is a director of Humanitarian Assistance and Disaster Relief Research Centre and a senior lecturer at Faculty of Defence Studies and Management, National Defence University of Malaysia (NDUM). Prior to her NDUM secondment in 2018, she serves the Royal Malaysian Navy for 25 years as a naval officer. Haliza Zahari holds a PhD in Technology, Operation and Logistics from Universiti Utara Malaysia. Her research interests are Defence Logistics, Humanitarian Logistics, Supply Chain Management and Public Procurement. She is an active member of Chartered International of Logistics and Transportation.

Ruzaidin Mohammed Zain is a part-time lecturer at the National Defence University of Malaysia (NDUM) and is currently pursuing a doctorate in Humanitarian Logistics. He previously served as a graduate research assistant at NDUM. With 13 years of experience as a naval officer in the Royal Malaysian Navy, Ruzaidin brings a wealth of practical knowledge to his academic role. He holds a master's degree in Project Management from the University of Malaya and has a keen interest in Defence Logistics, Humanitarian Supply Chain Management and Project Management. His current research focuses on humanitarian logistics, with a particular emphasis on information sharing for inter-agency coordination in disaster management.

Najuwa Mohd Nasir is a researcher and educator at Universiti Malaysia Perlis (UniMAP), affiliated with the Faculty of Business and Communication. She specializes in Technology Management and Resource Management in Logistics. Ms. Najuwa holds a Bachelor's and a Master's Degree in Technology Management from Universiti Utara Malaysia (UUM). She is currently pursuing advanced studies in Resource Management in Logistics at National Defence University of Malaysia (NDUM). Her research focuses on refining resource management practices within logistics. Additionally, she collaborates with various agencies on humanitarian logistics research, contributing to disaster relief and community support initiatives.

Nur Hidayah Azhar is an accomplished author with a Master's degree from National Defence University of Malaysia and has experience as a former graduate research assistant. She is also a graduate of Bachelor's Degree in Security and Defense Management with a first class honours from the National Defence University of Malaysia (NDUM). She is very interested in the research areas of Logistics and Transportation and Supply Chain Management.

A Thematic Review on Critical Success Factors in Humanitarian Logistics

3

Sharina Osman and Muna Norkhairunnisak Ustadi

Abstract

Humanitarian logistics plays a pivotal role in disaster response and relief operations, ensuring the effective and efficient delivery of aid to affected populations. This comprehensive thematic review synthesises existing literature on critical success factors (CSFs) in humanitarian logistics, aiming to identify key themes and factors that contribute to successful outcomes in humanitarian logistics operations. Drawing from interdisciplinary research in logistics, supply chain management, and disaster management, this paper identifies the key factors in humanitarian logistics operations including the effective coordination and collaboration, information and communication integration and management, flexibility and adaptability, risk management and resilience, and stakeholder engagement and capacity building.

Keywords

Humanitarian logistics · Critical success factors · Disaster relief · Supply chain management · Literature review

S. Osman (✉) · M. N. Ustadi
Business School, Universiti Kuala Lumpur, Kuala Lumpur, Malaysia
e-mail: sharina@unikl.edu.my

3.1 Introduction

Humanitarian logistics plays a vital role in addressing the needs of individuals and communities impacted by disasters and emergencies. It is an integral part of disaster response and relief operations, involving the planning, implementation, and management of goods, services, and information flow (Tatham and Kovács 2010). The effectiveness of humanitarian logistics depends on various factors, including stakeholder engagement, risk management, information management, flexibility, and coordination (Pettit et al. 2010). This thematic review aims to consolidate existing research on the critical success factors in humanitarian logistics, providing a comprehensive overview of what contributes to successful humanitarian operations.

Humanitarian logistics is a crucial discipline focused on efficiently organising and delivering aid during emergencies and disasters. It specialises in the prompt and effective distribution of essential resources such as food, water, shelter, and medical supplies to affected communities. The pivotal role of humanitarian logistics in mitigating disaster impacts, saving lives, and aiding recovery underscores its critical importance (Paciarotti et al. (2021)).

The importance of strong and efficient humanitarian logistics networks has become increasingly evident due to the rise in frequency

N. A. A. Rahman et al. (eds.), *Supply Chain in Humanitarian Operations*, Sustainable Development Goals Series,
https://doi.org/10.1007/978-981-96-4426-1_3

and intensity of both natural and man-made disasters in recent years. The persistent refugee crisis and the 2023 earthquakes in Turkey and Syria highlight the vital role logistics play in disaster management and recovery. This was reported by the United Nations Office for the Coordination of Humanitarian Affairs (OCHA) in 2023. Effective humanitarian logistics is crucial for the swift delivery of aid to those in need, which minimises suffering and aids in quicker recovery.

Technological progress and data analytics have greatly enhanced the humanitarian logistics field. The use of unmanned aerial vehicles (UAVs), or drones, to transport essential medical supplies and food to hard-to-reach areas has dramatically changed the landscape of humanitarian aid (Balasubramanian, Shukla et al. 2021). Additionally, blockchain technology is being explored as a way to increase supply chain transparency and traceability, ensuring that aid reaches its intended recipients securely and promptly, without any unauthorised interference (Kshetri 2021). These technologies are significantly boosting the efficiency and effectiveness of humanitarian logistics, transforming how it operates.

Despite advancements, humanitarian logistics continues to confront many challenges. The unpredictability of disasters, logistical limitations, scarce resources, and intricate operational settings present considerable obstacles. The COVID-19 pandemic underscored the importance of having logistics systems that are adaptable and sturdy, equipped to manage a wide range of health emergencies and concurrent crises (Banomyong et al. 2021). The pandemic placed immense pressure on worldwide supply chains, emphasising the critical need for humanitarian organisations to be ready, flexible, and well-coordinated.

Humanitarian logistics play pivotal role in disaster response and management. Its significance extends beyond simply transporting goods; it involves managing the entire supply chain to ensure that aid reaches those in need promptly and efficiently. To grasp the importance of humanitarian logistics, consider these key aspects:

1. Timeliness and Efficiency in Issue Resolution

Immediate action is vital following a disaster. The World Food Programme (WFP) was key in quickly distributing food to millions impacted by the 2010 Haiti earthquake. Despite facing logistical hurdles, the WFP's swift action averted a major famine disaster (WFP 2010). The disaster's effects were greatly reduced due to the rapid mobilisation and effective aid delivery by humanitarian organisations (Yáñez-Sandivari, L. et al. 2021). Timely aid can alleviate suffering and save lives. Efficient humanitarian logistics ensure the quick delivery of vital supplies, such as food, water, medical aid, and shelter, to communities in crisis.

2. Efficient Allocation of Resource Allocation

Efficient distribution and utilisation of finite resources are pivotal in humanitarian logistics, as highlighted by Kovács and Falagara, in 2021. This becomes crucial in disaster-stricken areas where resources are limited. For example, the International Federation of Red Cross and Red Crescent Societies (IFRC) managed logistics adeptly following the 2004 Indian Ocean tsunami, ensuring effective resource allocation and reaching a significant number of people across various countries (IFRC 2005). Effective logistics systems aid in fair distribution of assistance, cost savings, and minimising waste. The Malaysian Armed Forces (ATM) are instrumental in disaster response, using their logistical strengths to deliver relief promptly. In the 2018 Borneo floods, the ATM deployed assets like helicopters and vessels to deliver medical aid and food to isolated areas, ensuring support reached all communities in need (The Star 2018). Optimising this process is essential to improve the impact of every dollar spent on humanitarian aid.

3. Collaboration and Coordination

The effective collaboration between the United Nations, local administrations, and various NGOs enabled the swift and smooth delivery of aid during the 2015 Nepal earthquake, even amidst difficult and rugged terrain. This coordination proved crucial in reaching remote and otherwise inaccessible areas (UNOCHA 2015). The success of humanitarian logistics hinges on the cooperation and coordination among diverse stakeholders, such as government bodies, non-governmental organisations (NGOs), international groups, and the business sector. Efficient logistics operations facilitate the harmonious merging of efforts from multiple organisations, resulting in a more unified and impactful response (Negi 2022).

4. The Ability to Readily Adapt and Transform in Response to a Variety of Circumstances and Situations

To effectively tackle the dynamic and sometimes unpredictable nature of disasters, humanitarian logistics must be both adaptable and flexible. Unlike commercial logistics, humanitarian logistics needs to be ready to function in uncertain environments due to variable supply and demand conditions (Kumar P. et al. 2024). Swift adaptation to changing situations is crucial to ensure uninterrupted aid delivery. The COVID-19 pandemic highlighted the need for flexible logistics systems. Organisations like the World Health Organization (WHO) and various non-governmental organisations (NGOs) had to quickly adapt to new challenges, such as travel bans and lockdowns, to deliver medical supplies and vaccines worldwide (WHO 2021). In response to the pandemic, the Malaysian government founded the National Disaster Management Agency (NADMA) to streamline coordination between different agencies and NGOs. This joint effort was key in efficiently distributing personal protective equipment (PPE), medical resources, and food assistance to at-risk groups and frontline workers across the country (NADMA 2020).

5. Improvement of Capabilities and the Development of Tangible Structures

The World Food Programme's ongoing efforts to build local supply chain infrastructure and train local workers have bolstered the region's ability to tackle recurring droughts and food scarcities in East Africa. These initiatives have significantly strengthened resilience and food security (WFP 2019). Furthermore, investing in humanitarian logistics enhances the long-term resilience of impacted areas. The local capacity and infrastructure not only improve the speed of disaster response but also strengthen community defences against future crises. Training local personnel, building infrastructure, and creating strong supply networks are vital for sustainable development. In Malaysia, annual monsoon seasons bring frequent floods. Humanitarian organisations like MERCY Malaysia have developed adaptable logistics strategies to efficiently distribute relief items and deploy rapid response teams. This adaptability ensures that aid is delivered swiftly and effectively, especially when facing unexpected challenges (MERCY Malaysia 2020).

6. Humanitarian Principles

Humanitarian logistics is driven by the principles of impartiality, neutrality, independence, and humanity. The International Federation of Red Cross and Red Crescent Societies (IFRC) adheres strictly to these principles in conflict zones like Syria, ensuring that aid reaches the most vulnerable without regard to their stance in the conflict, thus building trust and cooperation among all involved (IFRC, 2020). These principles mandate that aid is allocated based on need alone, free from discrimination and unrelated to political or other associations. The Malaysian Red Crescent Society follows these same principles. During the 2015 Sabah earthquake, the MRCS ensured that assistance was provided according to the needs of all affected communities, regardless of their socio-economic status or location, promoting trust and cooperation in the relief process (MRCS, 2015). Establishing an ethical foundation is crucial for sustaining trust and cooperation in humanitarian endeavours.

3.2 Systematic Review Methodology

Thematic analysis is utilised in the review to identify recurring themes and patterns within the literature (Braun and Clarke 2012). To ensure a comprehensive and systematic approach, the journal search was conducted in distinct phases:

Phase 1: Planning and Preliminary Search

- Objective Setting: Define the review's objectives and scope, focusing particularly on the critical success factors of humanitarian logistics operations.
- Preliminary Search: Conduct a preliminary search to ascertain relevant keywords, search terms, and potential sources. The search strategy involved a mix of keywords and controlled vocabulary terms, such as "disaster relief," "critical success factors," "supply chain management," "risk management," and "humanitarian logistics," These terms were selected to cover a wide range of studies focusing on different aspects of humanitarian logistics operations. This phase aids in refining the search strategy and identifying potential databases and journals for the review.

Phase 2: Comprehensive Database Search

- Database Selection: Choose relevant academic databases, including Scopus, Google Scholar, and PubMed, known for their extensive coverage of scholarly literature across various disciplines, ensuring a wide-ranging and inclusive search.
- Search Execution: Execute an exhaustive search in each chosen database using the predetermined keywords and search terms. Enhance the retrieval of relevant studies by using a combination of keywords and Boolean operators.

Phase 3: Initial Screening

- Title and Abstract Screening: Evaluate the relevance of the retrieved articles to the research topic by reviewing their titles and abstracts. Articles that do not satisfy the inclusion criteria should be excluded.
- Inclusion and Exclusion Criteria: Utilise predetermined inclusion and exclusion criteria to eliminate studies that are irrelevant. In order to guarantee the currency and relevance of the data, articles published in peer-reviewed journals, books, and conference proceedings between 2010 and 2024 were chosen for inclusion in the review. Inclusion criteria prioritised empirical data, peer-reviewed articles, and English-language studies, while exclusion criteria prioritised studies that lacked primary data and non-English articles.

Phase 4: Comprehensive Review

- Full-Text Retrieval: Retrieve the full-text versions of the articles that successfully navigated the initial screening phase.
- Comprehensive Evaluation: Conduct a comprehensive evaluation of the full texts to determine their eligibility for inclusion in the systematic review. To guarantee that only pertinent and high-quality studies are incorporated, strictly enforce the inclusion and exclusion criteria.

Phase 5: Quality Assessment and Data Extraction

- Data Extraction: Systematically extract pertinent data from the selected studies, such as the study design, sample size, methodologies employed, and identified critical success factors.
- Quality Assessment: Utilise standardised tools or criteria to evaluate the reliability and quality of the studies that have been included. This phase guarantees the validity and robustness of the findings synthesised in the review.

Phase 6: Thematic Analysis and Data Synthesis

- Thematic Analysis: Perform thematic analysis to identify recurring themes and patterns in the selected studies. This entails the systematic examination of the literature by classifying and categorising data related to key success factors.
- Theme Synthesis: Develop overarching themes that represent the key success factors in humanitarian logistics operations by

synthesising the extracted data. This synthesis process offers a thorough comprehension of the variables that affect the efficiency and effectiveness of logistics operations in humanitarian environments.

3.3 Advantages of the Methodology

- Comprehensive Search Strategy: The review endeavoured to encompass a diverse array of studies and perspectives on the subject by employing multiple databases and meticulously chosen keywords.
- Transparent Selection Criteria: In order to guarantee the quality and relevance of the studies included, explicit inclusion and exclusion criteria were implemented.
- Systematic Data Extraction and Synthesis: The consistent themes and patterns were identified through the systematic approach to data extraction and thematic synthesis, which facilitated a structured analysis of the literature.

3.4 Limitations of the Methodology

- Language Bias: The search may have been incomplete if it was limited to articles published in English, which could have resulted in the exclusion of pertinent studies published in other languages.
- Publication Bias: The review focused predominantly on peer-reviewed journal articles, which may have overlooked valuable insights from other sources, such as grey literature or reports.

The methodology described above offers a structured and rigorous approach to conducting a systematic review of the critical success factors in humanitarian logistics operations. The objective of this review was to enhance comprehension of the factors that influence the effectiveness of humanitarian logistics and to offer insights for practice and future research by conducting a systematic search, selection, and synthesis of pertinent literature.

3.5 Key Success Factors in Humanitarian Logistics Operations

Theme 1: Coordination and Collaboration

Due to the urgency and complexity of disaster relief efforts, the success of humanitarian logistics operations is contingent upon the effective coordination and collaboration of stakeholders (Thomas and Kopczak 2005). The harmonisation of activities among various organisations involved in humanitarian response, such as government agencies, non-governmental organisations (NGOs), and international organisations, is a process known as coordination. It is the process that aligns the efforts of different organisations involved in humanitarian responses (Pettit and Beresford 2009). Coordination among humanitarian actors is essential for harmonising activities and ensuring that resources are used efficiently. Collaboration facilitates the efficient and timely delivery of aid to affected populations by promoting information sharing, resource pooling, and collaborative decision-making (Pettit et al. 2010). For instance, Smith et al. (2018) emphasise that the efficiency of logistics processes and the optimisation of resource allocation can be achieved through the establishment of effective partnerships between these entities. Successful coordination necessitates consistent communication, shared information systems, and joint planning.

The United Nations' Logistics Cluster was instrumental in the coordination of the activities of numerous humanitarian organisations during the 2010 Haiti earthquake. The cluster enhanced the efficiency of logistics operations by offering a platform for information exchange and collaborative decision-making, thereby guaranteeing the timely delivery of aid to the affected regions. The 2010 Haiti earthquake was a catastrophic event that caused extensive destruction and a significant loss of life. The United Nations World Food Programme led the Logistics Cluster, which functioned as a coordination mechanism to oversee the logistics of humanitarian assistance. The cluster facilitated the pooling of resources, prevented the duplication of efforts, and guaranteed the delivery of critical supplies, including food, water, and medical apparatus, to the most critical locations. The overall response effort was improved.

Theme 2: Management and Integration of Information and Communication

In humanitarian logistics, the allocation of resources and the formulation of decisions are contingent upon the availability of precise and expeditious information (Tomasini and Van Wassenhove, 2009). Information management systems enable organisations to effectively respond to changing conditions and priorities by facilitating the collection, analysis, and dissemination of data on inventory levels, transportation routes, and requirements assessment (Altay and Labonte 2011). The integration of ICT improves the efficiency and effectiveness of humanitarian logistics by supporting decision-making processes, enhancing communication, and providing real-time data. Information sharing platforms and technologies, including satellite imagery and geographic information systems (GIS), are essential for improving situational awareness and coordination among stakeholders (Balcik et al. 2016). In the same way, mobile applications facilitate the sharing of real-time data and the improved coordination of relief efforts. Kovács and Spens (2019) underscore the importance of ICT in its ability to improve situational awareness, monitor supply chain performance, and trace the movement of goods. ICT tools facilitate the rapid response of humanitarian organisations to evolving circumstances and the formulation of well-informed decisions that are based on precise and current information.

UAVs (drones) outfitted with high-resolution cameras were utilised during the 2015 Nepal earthquake to acquire valuable aerial imagery for the purpose of evaluating damage and organising logistics operations. This technology enabled humanitarian organisations to prioritise the delivery of aid and identify inaccessible areas. The Nepal earthquake resulted in significant infrastructure damage, rendering numerous regions inaccessible by road. Drones were employed by humanitarian organisations to acquire high-resolution aerial images of the afflicted regions. The logistics teams were able to identify the most afflicted areas and plan the most efficient routes for delivering aid as a result of the detailed information provided by these images on the extent of damage. The real-time data from drones facilitated the rapid and precise decision-making process, thereby guaranteeing that aid was delivered to the individuals who required it most.

Theme 3: Adaptability and Flexibility

Logistics systems must be adaptable and flexible in response to evolving circumstances due to the dynamic nature of humanitarian crises (Tatham and Christopher, 2018). In response to changing requirements, constraints, and opportunities, flexibility allows organisations to modify their strategies, processes, and resources (Tatham and Christopher, 2018). In order to improve the resilience and responsiveness of humanitarian operations, adaptive logistics systems anticipate and mitigate risks, minimise disruptions, and optimise the utilisation of available resources (Kovács and Spens 2007). Responses to humanitarian crises necessitate adaptability and flexibility due to their dynamic and frequently unpredictable nature. Supply chain design flexibility and responsiveness to changing conditions (e.g., infrastructure damage, weather) facilitate the rapid deployment of resources and the modification of distribution strategies. Pettit et al. (2020) contend that the vagaries and variability that are inherent in disaster scenarios can be more effectively managed through the implementation of flexible supply chains. This encompasses the establishment of contingency plans, the diversification of supply sources, and the preservation of buffer stocks to guarantee the continuity of operations. During the 2014

Ebola outbreak in West Africa, humanitarian organisations were compelled to promptly modify their logistics strategies in order to accommodate quarantine zones, mobility restrictions, and enhanced security measures. The crisis was effectively managed by the capacity to promptly adjust supply routes and relocate resources.

The 2014 Ebola outbreak in West Africa presented a multitude of logistical challenges, such as the necessity for specialised medical supplies, restricted movements, and quarantined areas. Humanitarian organisations were required to be highly adaptable, frequently revising their logistics plans to address these challenges. For instance, they implemented air bridges to transport supplies to remote regions, established mobile treatment units, and established decentralised supply centres to guarantee that resources could be rapidly mobilised and redirected as required. This adaptability was crucial in the management of the contagion and the provision of timely medical care to the affected populations.

Theme 4: Resilience and Risk Management

It is imperative to implement effective risk management in order to mitigate threats and vulnerabilities in humanitarian logistics operations (Pettit et al. 2010). Organisations are able to identify and prioritise risks associated with environmental hazards, security threats, and supply chain disruptions through the use of risk assessment techniques such as scenario planning, probability analysis, and vulnerability mapping (Sun et al. 2022). In order to improve the resilience of logistics networks, risk mitigation strategies include diversifying suppliers, pre-positioning emergency supplies, and implementing contingency plans (Pettit and Beresford 2009). Humanitarian logistics operations necessitate resilience and risk management. Organisations enhance their operational resilience by conducting pre-emptive risk assessments, scenario planning, and the development of contingency plans, which enable them to anticipate and resolve prospective challenges. Van Wassenhove (2012) contends that resilient supply chains are more capable of withstanding disruptions and recovering from disturbances more rapidly. This entails the identification of potential risks, the development of mitigation strategies, and the investment in infrastructure and resources to support recovery efforts. The International Federation of Red Cross and Red Crescent Societies (IFRC) implements a comprehensive risk management framework that encompasses the pre-positioning of relief supplies, contingency planning, and risk assessments. This method has allowed the organisation to preserve operational continuity in the face of a variety of crises, such as conflict and natural disasters. The IFRC's risk management framework is intended to fortify the organisation's resilience in the presence of a variety of crises. This framework entails the execution of comprehensive risk assessments to detect potential vulnerabilities and hazards. The IFRC formulates contingency plans that specify the actions to be taken in response to various circumstances, based on these assessments. Additionally, the International Federation of Red Cross and Red Crescent Societies (IFRC) prepositions relief supplies in strategic locations worldwide to guarantee that critical resources are rapidly accessible for deployment. This proactive approach allows the IFRC to effectively respond to emergencies and maintain continuity in its operations, thereby minimising the impact on affected populations.

Theme 5: Capacity Building and Stakeholder Engagement

It is imperative to establish trust, legitimacy, and acceptability in humanitarian logistics operations by engaging with local communities, governments, and other stakeholders (Pettit and Beresford 2009). Stakeholder engagement ensures that humanitarian assistance is customised to the preferences and requirements of impacted populations, thereby fostering accountability, cultural sensitivity, and participatory decision-making (Tatham and Christopher, 2018). The involvement of stakeholders in the planning, implementation, and evaluation of humanitarian response efforts is facilitated by collaborative approaches, such as community-based logistics and social network analysis (Altay and Labonte 2014).

It is imperative to establish the capacity of staff and volunteers through ongoing training programmes in order to ensure the success of humanitarian logistics operations. Personnel are capable of fulfilling their obligations more efficiently through training that enhances their abilities in logistics management, emergency response, and cultural sensitivity. Kopczak and Thomas (2021) emphasise the significance of training programmes in improving the preparedness and responsiveness of humanitarian organisations. In order to guarantee sustainability, capacity building also entails the enhancement of local collaborators' capabilities and the cultivation of local expertise.

The World Food Programme (WFP) conducts training programmes on a regular basis for its staff and partners, covering topics such as emergency logistics, warehouse management, and supply chain management. The organisation's capacity to respond to crises with efficiency has been significantly improved by these programmes.

A highly qualified workforce is necessary to effectively manage logistics in the diverse and frequently challenging environments in which the World Food Programme (WFP) operates. The World Food Programme (WFP) implements comprehensive training programmes that encompass a wide range of logistics and supply chain management topics in order to guarantee readiness. These programmes encompass seminars on coordination and communication skills, technical training on warehousing and transportation, and simulations of emergency scenarios. By investing in the ongoing development of its staff and partners, the World Food Programme (WFP) improves its operational capacity and guarantees that it can rapidly and effectively address emergencies.

3.6 Conclusion

This thematic review identifies and synthesises critical success factors that are essential for enhancing the efficiency and efficacy of humanitarian logistics. The success of humanitarian logistics operations is contingent upon the effective coordination and collaboration, information and communication integration and management, flexibility and adaptability, risk management and resilience, and stakeholder engagement and capacity building. Humanitarian organisations can reduce the impact on affected populations and contribute to the broader objectives of sustainable development and resilience building by improving their ability to respond effectively to disasters and emergencies by addressing these Critical Success Factors.

Humanitarian logistics operations are essential for the provision of expeditious and effective assistance to populations that have been impacted by emergencies and disasters. The efficiency and effectiveness of these operations are guaranteed by the identification of numerous critical success factors in the systematic review conducted here. These factors include adaptability and flexibility, coordination and collaboration, ICT integration, capacity development and training, and risk management and resilience.

It is essential to ensure that a diverse array of stakeholders, such as NGOs, government agencies, international organisations, and local communities, collaborate effectively. These entities can optimise the utilisation of available resources, address critical gaps, and prevent duplication by aligning efforts and resources. The incorporation of information and communication technology is a game-changer in humanitarian logistics. Situational awareness and decision-making are considerably enhanced by real-time data sharing and enhanced communication, which are facilitated by technologies such as GIS, satellite imagery, and mobile applications. A high level of adaptability and flexibility is required due to the unpredictable and dynamic nature of humanitarian crises. Supply chains must be engineered to be adaptable to evolving circumstances, enabling the rapid deployment of resources and the modification of distribution strategies. In order to improve the capacity of humanitarian organisations, it is imperative to invest in the development of personnel through ongoing training programmes. Training in logistics management, emergency response, and cultural sensitivity guarantees that volunteers and personnel are adequately equipped to fulfil their responsibilities. Resilience planning and effective risk management are essential for maintaining the continuity of operations and mitigating disruptions. Scenario planning, pre-emptive risk assessments, and contingency plans provide organisations with the capacity to anticipate challenges and improve their response to crises.

Humanitarian organisations can enhance their preparedness and responsiveness, resulting in improved outcomes for affected populations, by comprehending and implementing these factors. These strategies are essential for the advancement of humanitarian logistics and the efficient and effective delivery of aid to those in need. Their continuous development and refinement, which are based on empirical research and practical experience, are essential.

In summary, the success of humanitarian logistics operations is contingent upon a multifaceted approach. These components, when combined, establish a strong framework that enables humanitarian organisations to provide assistance in a timely and efficient manner, regardless of the most unpredictable and difficult circumstances. In order to identify new best practices and further enhance the capacity of humanitarian logistics to save lives and alleviate suffering during times of crisis, ongoing research and collaboration will be essential as the field continues to evolve.

References:

Altay NEZIH, Labonte MELISSA (2011) Humanitarian logistics and the cluster approach. Humanitarian Logistics. 97–114

Balasubramanian S, Shukla M, Ramesh A (2021) The role of drones in humanitarian logistics: An overview. J Hum Logist Supply Chain Manag 11(2):156–172

Balcik B, Bozkir CDC, Kundakcioglu OE (2016) A literature review on inventory management in humanitarian supply chains. Surv Oper Res Manag Sci 21(2):101–116

Banomyong R, Varadejsatitwong P, Oloruntoba R (2021) Managing humanitarian logistics during COVID-19: The case of Thailand. Int J Logist Manag 32(2):471–488

Braun V, Clarke V (2012) Thematic analysis. American Psychological Association

Jabatan Kebajikan Masyarakat (2015) Kelantan floods report 2014. Retrieved from https://www.jkm.gov.my

Kovács G, Falagara Sigala I (2021) Lessons learned from humanitarian logistics to manage supply chain disruptions. J. Supply Chain Manag. 57(1):41–49

Kovács G, Spens KM (2007) Humanitarian logistics in disaster relief operations. Int J Phys Distrib Logist Manag 37(2):99–114

Kovács G, Spens KM (2019) Humanitarian logistics revisited: The role of logistics service providers. J. Hum. Logist. Supply Chain Manag. 9(1):114–131

Kshetri N (2021) Blockchain in humanitarian supply chains: Overview and case studies. Technol Forecast Soc Chang 170:120951

Kumar P, Singh RK, Shahgholian A (2024) Learnings from COVID-19 for managing humanitarian supply chains: systematic literature review and future research directions. Ann Oper Res 335(3):899–935

Malaysian red crescent society (MRCS) (2015) Sabah earthquake response overview. Retrieved from https://www.redcrescent.org.my

MERCY Malaysia (2020) Annual report 2020. Retrieved from https://www.mercy.org.my

National Disaster Management agency (NADMA) (2020) COVID-19 Response report. Retrieved from https://www.nadma.gov.my

Negi S (2022) Humanitarian logistics challenges in disaster relief operations: A humanitarian organisations' perspective. J Transp Supply Chain Manag 16:691

Paciarotti C, Piotrowicz WD, Fenton G (2021) Humanitarian logistics and supply chain standards. Literature review and view from practice. J. Hum. Logist. Supply Chain Management. 11(3):550–573

Pettit SJ, Beresford AKC (2009) Critical success factors for the effective humanitarian relief supply chain management in disasters. Int J Prod Econ 126(1):55–65

Pettit TJ, Fiksel J, Croxton KL (2010) Ensuring supply chain resilience: Development of a conceptual framework. J Bus Logist 31(1):1–21

Pettit S, Beresford A, Mina S (2020) Critical success factors in humanitarian logistics: A systematic review. Int. J. Phys. Distrib. Logist. Manag. 50(4):343–367

Sarawak. Retrieved from https://www.my.undp.org

Smith T, Adhikari R, Sarker S (2018) Coordination in humanitarian logistics: Challenges and opportunities. Int. J. Prod. Econ. 199:84–95

Sun H, Li J, Wang T, Xue Y (2022) A novel scenario-based robust bi-objective optimization model for humanitarian logistics network under risk of disruptions. Transportation Research Part e: Logistics and Transportation Review 157:102578

Tatham PH, Kovács G (2010) The application of military logistics to humanitarian logistics. Int J Phys Distrib Logist Manag 40(8/9):639–656
Tatham P, Christopher M (eds) (2018) Humanitarian logistics: Meeting the challenge of preparing for and responding to disasters. Kogan Page Publishers
The Star (2018) ATM's role in Borneo flood relief. Retrieved from https://www.thestar.com.my
Thomas A, Kopczak LR (2005) From logistics to supply chain management: The path forward in the humanitarian sector. Fritz Institute White Paper Series 1(1):1–21
Thomas A, Kopczak LR (2021) Humanitarian supply chain management: A review and future research agenda. J. Hum.itarian Logist. Supply Chain. Manag. 11(3):342–360
Tomasini RM, Van Wassenhove LN (2009) From preparedness to partnerships: case study research on humanitarian logistics Int Trans Oper Res 16(5):549-559
United Nations Development Programme (2019) Infrastructure improvement projects in Sabah and United Nations office for the coordination of humanitarian affairs (OCHA) (2023). Global humanitarian overview 2023. Retrieved from https://www.unocha.org
United Nations office for the coordination of humanitarian affairs (OCHA) (2023) Global humanitarian overview 2023. Retrieved from https://www.unocha.org
UNOCHA (2015) Nepal earthquake 2015: An overview of the humanitarian response. Retrieved from https://www.unocha.org
Van Wassenhove LN (2012) Humanitarian aid logistics: Supply chain management in high gear. J. Oper. Res. Soc. 63(1):31–49
WFP (2010) Haiti earthquake response. Retrieved from https://www.wfp.org
WFP (2019) East Africa: Building resilience through local capacity building. Retrieved from https://www.wfp.org
Yáñez-Sandivari L, Cortés CE, Rey PA (2021) Humanitarian logistics and emergencies management: New perspectives to a sociotechnical problem and its optimization approach management. Int J Disaster Risk Reduct 52:101952

Dr. Sharina Osman is a senior lecturer at the Management & Entrepreneurship section and a deputy dean of Student Development & Campus Lifestyle of Universiti Kuala Lumpur Business School, Malaysia. She earned her PhD degree in Management from the University of Exeter, United Kingdom; Master of Human Resource Management and BBA (Marketing) Honours from Universiti Utara Malaysia (UUM), Malaysia. She has more than ten years of teaching experience at the university level. Identity, Image and Reputation, Organisational Culture, Human Resource Development, Tourism, Tourism Destination Identity & Branding, Participative Tourism Development, and e-Tourism. Dr Osman carries an array of experience ranging from Human Resource to Marketing (Image & Branding) and Tourism. She is also the Editorial Advisory Board for several journals and has been invited as a Keynote or Guest Speaker in seminars. Being in an academic and in the research environment, coupled with vast hands-on experience in the industry, she helps organisations improve competency and capacity building through trainings and consultations. She believes in interacting and experiential learning and has conducted many workshops in different sectors, including government and private at both local and international.

Ts. Dr. Muna Norkhairunnisak Ustadi holds the position of a senior lecturer and is responsible for overseeing the Quality Assurance department at Universiti Kuala Lumpur Business School. She possesses extensive knowledge and expertise in supply chain management, logistics, and transportation. She possesses a wealth of expertise in the logistics sector, having worked with major logistics corporations. She regularly engages in lectures and actively participates in research fields related to supply chain management, logistics, and transportation. She possesses a Doctorate in Technology Management from Universiti Tun Hussein Onn (UTHM), a Master's degree in Business Administration from Universiti Teknologi Mara (UiTM), and a Bachelor's degree in International Business from Univerisiti Utara Malaysia. She has the prestigious title of Chartered Member of the Chartered Institute of Logistics and Transport Malaysia (CILTM). In addition, she obtained certification as a professional technologist from the Malaysia Board of Technology.

Part II
Technology Trends in Humanitarian Logistics

Innovation Adoption in Humanitarian Logistics: A Framework for Evaluating the Implementation of Cutting-Edge Technologies in Relief Operations

4

Aisyah Othman and Zulkarnian Ahmad

Abstract

Innovation adoption in humanitarian logistics is crucial for enhancing the efficiency and effectiveness of relief operations in response to disasters and crises. This study presents a comprehensive framework for evaluating the implementation of cutting-edge technologies in humanitarian logistics, focusing on the unique challenges and opportunities within this sector. By integrating theoretical perspectives on innovation adoption with practical insights from field operations, the framework aims to provide a robust tool for assessing the readiness, feasibility, and impact of new technologies in diverse humanitarian contexts. The research highlights key factors influencing the adoption process, such as organizational readiness, stakeholder engagement, and the adaptability of technology to the unpredictable and resource-constrained environments typical of humanitarian efforts. Through this framework, the study seeks to guide policymakers, practitioners, and researchers in making informed decisions about integrating innovative solutions into humanitarian logistics, ultimately improving the timeliness and effectiveness of aid delivery to those in need.

Keywords

Innovation adoption · Technology adoption · Humanitarian logistics · Supply chain management and operation

A. Othman (✉) · Z. Ahmad
Faculty of Business and Management, Akademi Laut Malaysia, Melaka, Malaysia
e-mail: aisyah.othman@alam.edu.my

Z. Ahmad
e-mail: zulkarnian.ahmad@alam.edu.my

4.1 Introduction

Overview of Humanitarian Logistics and the Need for Innovation

The necessity for innovation in humanitarian logistics has become increasingly obvious due to the growing scale and complexity of contemporary humanitarian crises. Traditional logistical methods often fall short in effectively responding to these challenges as such, the adoption of innovative solutions, particularly those driven by technological advancements, is critical for enhancing the capacity of humanitarian organizations to deliver aid more efficiently and effectively. This involves integrating new tools, processes, and strategies that can optimize the coordination and performance of logistics operations, thereby extending the reach and impact of humanitarian. Consequently, the need for

N. A. A. Rahman et al. (eds.), *Supply Chain in Humanitarian Operations*, Sustainable Development Goals Series,
https://doi.org/10.1007/978-981-96-4426-1_4

innovation in humanitarian logistics transcends the goal of improving operational efficiency; it is essential for saving lives and reducing human suffering in the aftermath of disasters.

The Role of Technology in Addressing Humanitarian Challenges

The role of these elements are illustrated how technology is integral to overcoming the challenges faced in humanitarian logistics, ultimately improving the speed, efficiency, and impact of disaster response efforts.

Enhanced Communication and Coordination

Technology enables real-time communication and efficient collaboration among humanitarian actors, even in remote or disaster-stricken areas.

Supply Chain Optimization

AI and machine learning improve supply chain management by predicting demand, managing inventory, and planning distribution routes for timely aid delivery.

Drones and Robotics

Drones and robotics revolutionize supply delivery and search operations in hard-to-reach or dangerous areas, improving response times.

Blockchain for Transparency and Accountability

Blockchain ensures secure, transparent tracking of resources and funds, enhancing accountability in humanitarian operations.

Real-Time Data and Predictive Analytics

Real-time data and predictive analytics help organizations check situations, anticipate needs, and distribute resources effectively.

Geospatial Technologies and Mapping

Geospatial technologies like GIS provide right mapping and damage assessment for effective disaster response planning.

Mobile Technology and Crowdsourcing

Mobile technology and crowdsourcing empower affected communities to communicate needs and aid humanitarian response.

In conclusion, technology-driven innovation in humanitarian coordination is vital for enhancing coordination, efficiency, and impact, ultimately saving lives and alleviating human suffering in increasingly complex crisis situations.

4.2 Theoretical Foundations of Innovation Adoption

The theoretical foundations of innovation adoption explore the underlying principles and models that explain how and why organizations integrate new technologies into their operational frameworks and decision-making processes.

Diffusion of Innovations Theory

The Diffusion of Innovations Theory, developed by Everett Rogers (1962, 2003) explains how, why, and at what rate new ideas and technology spread within a culture. When applied to evaluating the implementation of cutting-edge technologies in relief operations, this theory can offer a structured approach to understanding the adoption process among various stakeholders, including governments, NGOs, and affected communities. Table 4.1 shows the Diffusion of Innovations Theory description.

When evaluating cutting-edge technologies in relief operations, the Diffusion of Innovations Theory can help identify potential barriers to adoption, assess the effectiveness of communication strategies, and understand the role of early adopters in promoting wider acceptance.

Technology Acceptance Model (TAM)

The Technology Acceptance Model (TAM) is a theoretical framework used to understand and predict user behavior in relation to technology

Table 4.1 The diffusion of innovations theory description

Key component	Description
Innovation	The new technology or method introduced in relief operations, such as advanced communication tools, drones, or data analytics platforms. The innovation must demonstrate a clear advantage over existing methods to be considered for adoption
Communication channels	The pathways through which information about the technology is disseminated among potential adopters. In relief operations, these channels might include training programs, demonstration projects, or collaborative platforms
Time	The pathways through which information about the technology is disseminated among potential adopters. In relief operations, these channels might include training programs, demonstration projects, or collaborative platforms
Social system	The network of individuals and organizations involved in relief efforts. The framework would analyze how social norms, organizational culture, and existing infrastructure impact the adoption of new technologies

adoption. Developed by Fred Davis in 1989, Fig. 4.1 is the theoretical framework of TAM, based on the Theory of Reasoned Action (TRA) and is specifically designed to explain and forecast how users come to accept and use a technology.

TAM has been widely applied across various industries to study the adoption of new technologies, including information systems, healthcare technologies, e-learning platforms, and more. It provides a robust framework for organizations to understand the factors that drive technology adoption among users, which can be critical for the successful implementation of new systems.

Theory of Planned Behavior and Technology

The Theory of Planned Behavior (TPB), developed by Icek Ajzen in 1991, is an extension of the Theory of Reasoned Action (TRA) and serves as a robust framework for understanding and predicting human behavior, particularly in the context of technology adoption. TPB posits that an individual's behavior directly influenced by their intention to perform the behavior, which represent these three key factors:

Attitude Toward the Behavior

This is how positively or negatively a person views using technology. If they believe the technology will benefit them, they are more likely to adopt it.

Subjective Norms

This involves the influence of others, such as peers or organizational culture, on a person's decision to use the technology. If people around them support the technology, they are more likely to adopt it.

Perceived Behavioral Control

This refers to the person's belief in their ability to use technology. If they feel capable and have the necessary resources, they are more likely to adopt the technology.

In short, TPB shows that positive attitudes, supportive social pressures, and confidence in usage lead to higher chances of adopting new technology.

Technology-Organization-Environment Theory

The Technology-Organization-Environment (TOE) Framework is a theoretical model that explains the factors influencing the adoption and implementation of technology within organizations.

Figure 4.2, developed by Tornatzky and Fleischer in 1990, is the illustration of the TOE Framework that identifies three critical contexts that impact technology adoption:

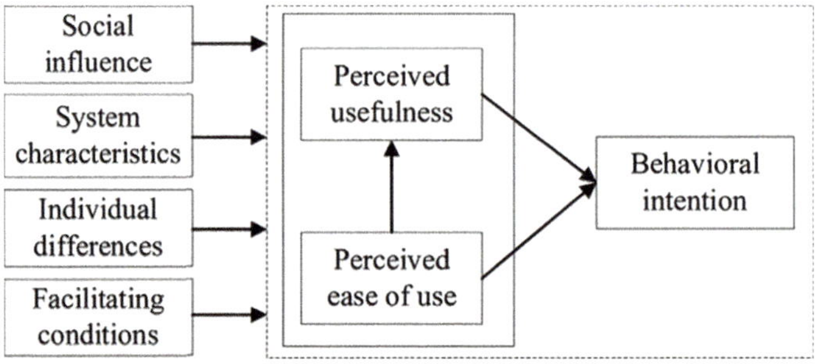

Fig. 4.1 The theoretical framework of the technology acceptance model (TAM)

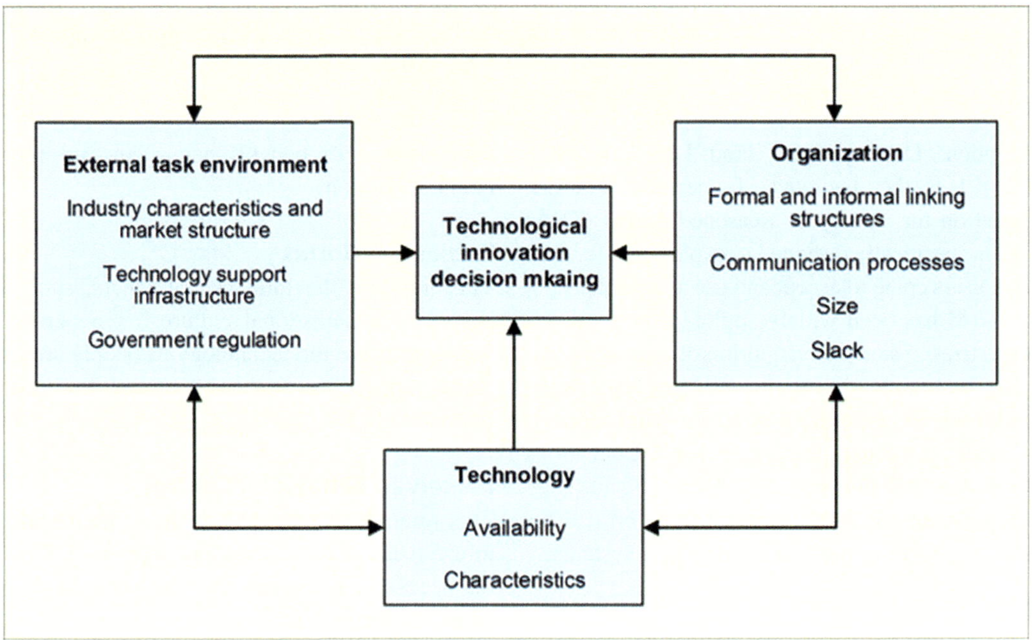

Fig. 4.2 Adapted from TOE adoption theoretical framework (Tornatzky et al. 1990)

The adoption and implementation of cutting-edge technologies in relief operations are significantly influenced by several factors, which can be effectively analyzed using the Technology-Organization-Environment (TOE) Framework. When evaluating the integration of advanced technologies in relief settings, it is essential to consider technological aspects such as compatibility with existing systems and the complexity of fresh solutions. Organizational factors, including the availability of resources, management support, and the organizational culture, are critical in determining the success of such implementations.

Additionally, external environmental factors, such as regulatory requirements, stakeholder pressures, and the broader socio-political context, play a crucial role in shaping the adoption process.

Innovation Resistance and Overcoming Barriers to Technology Adoption in Humanitarian Logistics

Innovation in humanitarian logistics can greatly enhance disaster response, but adopting new technologies often faces resistance due to organizational, cultural, and operational barriers. Risk aversion within organizations, driven by the mission-critical focus on saving lives, leads to a preference for established methods over unproven technologies. Cultural resistance arises from deeply ingrained practices and skepticism toward innovation, especially when it challenges traditional approaches. Operational barriers, such as limited infrastructure, resource constraints, and high costs, further hinder technology adoption.

To overcome these challenges, fostering a culture of innovation is essential. This involves encouraging experimentation, rewarding creative problem-solving, and strong leadership to champion modern technologies. Education and training are crucial for reducing resistance by equipping staff with the necessary skills and confidence to use new tools effectively.

Additionally, collaboration with technology providers and other stakeholders can help tailor solutions to the unique challenges of humanitarian operations and provide the resources needed for successful implementation. In conclusion, while resistance to innovation is a significant challenge in humanitarian logistics, addressing these barriers can enable organizations to harness the full potential of modern technologies and improve their disaster response capabilities.

4.3 Organizational and Strategic Perspectives

Innovation adoption in humanitarian logistics is crucial for enhancing the efficiency and effectiveness of relief operations. The key elements which can focus on the organizational and strategic perspectives that play a significant role in implementing cutting-edge technologies in humanitarian contexts.

Organizational Culture in the Adoption of Innovation

Organizational culture is critical in the adoption of innovation, particularly in humanitarian logistics. A culture that encourages experimentation, values learning, and is open to change is more likely to successfully integrate new technologies. In humanitarian organizations, where operations are often high-stakes and time-sensitive, fostering a culture that supports innovation can lead to more agile and responsive logistics processes. Resistance to change, on the other hand, can hinder the adoption of potentially life-saving innovations.

For example, Médecins Sans Frontières (MSF) has cultivated a culture that prioritizes innovation by embracing mobile health (mHealth) technologies to enhance patient care in remote areas. By fostering an environment where staff are encouraged to test and adapt new tools, MSF has been able to streamline their operations and improve their response times in crises.

Leadership in Driving Innovation Adoption

Leadership plays a pivotal role in driving the adoption of innovation within humanitarian organizations. Effective leaders advocate for new technologies and inspire and motivate their teams to embrace change. Leadership in this context involves creating a vision for how innovation can improve operations, securing the necessary resources, and guiding the organization through the transition process. Leaders must also manage the risks associated with adopting new technologies, ensuring that innovations are implemented in a way that enhances, rather than disrupts, humanitarian efforts.

For instance, during the 2014 Ebola outbreak in West Africa, the leadership of the World Health Organization (WHO) emphasized the importance of adopting innovative data collection tools. They introduced the use of mobile data platforms to track the spread of the virus in real time, which significantly improved the coordination of relief efforts. This example highlights how strong leadership can accelerate the adoption of critical innovations in emergency situations.

Strategic Alignment and Innovation

Strategic alignment refers to the degree to which an organization's innovation efforts are in sync with its overall goals and objectives. In humanitarian logistics, this means ensuring that the adoption of new technologies directly supports the mission of providing timely and effective relief to those in need. Strategic alignment helps organizations prioritize innovations that will have the most significant impact on their operations and ensures that resources are allocated efficiently. Without this alignment, even well-intentioned innovations may fail to deliver the expected benefits or could divert attention and resources from more critical areas.

For example, the United Nations World Food Programme (WFP) strategically aligned its use of blockchain technology to enhance food distribution. By aligning this innovation with their goal of reducing food insecurity, WFP has been able to streamline supply chains and ensure that aid reaches those in need more efficiently.

Dynamic Capabilities and Innovation

Dynamic capabilities refer to an organization's ability to adapt to changing environments by integrating, building, and reconfiguring internal and external competencies. In the context of humanitarian logistics, dynamic capabilities enable organizations to rapidly adopt and implement new technologies in response to evolving challenges in the field. This includes the ability to quickly scale up the use of a new logistics platform during a large-scale disaster or to modify existing processes to incorporate new data analytics tools. Organizations with strong dynamic capabilities are better positioned to leverage innovations to improve their responsiveness and effectiveness in delivering aid.

For example, the International Federation of Red Cross and Red Crescent Societies (IFRC) demonstrated dynamic capabilities by quickly integrating drone technology into their disaster response efforts. During the 2019 Cyclone Idai in Mozambique, IFRC used drones to assess damage and identify inaccessible areas, allowing them to adjust their logistics and improve the delivery of aid.

4.4 Institutional and Stakeholder Influences

Play a pivotal role in shaping organizational behavior and decision-making. Institutions, including regulatory bodies, professional associations, and cultural norms, establish the formal and informal rules that organizations must navigate. These institutions create a framework of expectations and standards that organizations are expected to adhere to, influencing everything from strategic choices to daily operations. Meanwhile, stakeholders such as customers, employees, investors, and the community exert their influence by holding organizations accountable to their interests and expectations. Balancing these often-competing demands requires organizations to be adaptive, ethical, and transparent, ensuring that they can sustain legitimacy and achieve long-term success within their operating environment.

The Functions

The function of institutions and stakeholders is integral to the governance and sustainability of organizations. Institutions provide the structural framework, encompassing laws, regulations, and societal norms, which dictate the rules of the game for organizations. They ensure that organizations operate within certain boundaries, promoting stability and consistency across industries. Stakeholders, on the other hand, serve as the voices of accountability and interest. They actively engage with organizations, influencing their strategies and actions through their demands, feedback, and support. Together, institutions and stakeholders function as the dual pillars that guide, regulate, and shape organizational behavior, ensuring that it aligns with broader societal goals and expectations while fostering trust and legitimacy.

4.5 Knowledge, Learning, and Capacity Building

The successful adoption of cutting-edge technologies in humanitarian logistics hinges on a robust foundation of knowledge, continuous learning, and capacity building. These elements are essential for organizations striving to enhance their operational efficiency, adaptability, and resilience in the face of complex humanitarian challenges.

Absorptive Capacity and the Adoption of Cutting-Edge Technologies in Humanitarian Logistics

Absorptive ability refers to an organization's ability to recognize the value of added information, assimilate it, and apply it to achieve strategic objectives. In the context of humanitarian logistics, this ability is crucial for the adoption of cutting-edge technologies that can revolutionize the way relief operations are conducted. Organizations with high absorptive ability are better equipped to integrate modern technologies into their existing frameworks, ensuring that innovations are not only adopted but also optimized to meet specific operational needs.

To build and enhance absorptive ability, humanitarian organizations must invest in continuous learning and development programs. These initiatives should focus on both technical skills and the ability to adapt to new methodologies and tools. Moreover, fostering a culture of innovation and openness to change within the organization can significantly enhance its absorptive ability. By doing so, humanitarian organizations can ensure that they are not merely passive recipients of technological advancements but active participants in shaping how these technologies are utilized in the field.

Human-Centered Design in the Development of Technological Solutions for Humanitarian Logistics

Human-centered design (HCD) is a problem-solving approach that involves the active participation of end-users in the design process to create solutions that are tailored to their needs. In humanitarian logistics, where the stakes are often high, employing HCD principles is vital to ensure that technological solutions are not only effective but also user-friendly and contextually proper.

The integration of HCD in the development of technological solutions for humanitarian logistics ensures that the end products align with the specific challenges and requirements of the users in the field. This approach involves iterative testing and feedback loops, enabling designers to refine technologies based on real-world conditions and user experiences. As a result, the solutions developed are more likely to be adopted and successfully implemented, as they directly address the needs and constraints faced by humanitarian workers.

Incorporating HCD into the innovation process also facilitates greater buy-in from stakeholders, as it shows a commitment to creating technologies that genuinely enhance their capabilities and effectiveness. By prioritizing the user experience, humanitarian organizations can drive the successful adoption of modern technologies and improve the overall impact of their relief operations.

Cross-Sector Collaboration in Technology Adoption for Humanitarian Logistics

The adoption of cutting-edge technologies in humanitarian logistics is often a complex process that requires collaboration across multiple sectors, including government agencies, non-governmental organizations (NGOs), academia, and the private sector. Cross-sector collaboration brings together diverse perspectives, expertise, and resources, enabling the development and implementation of more comprehensive and effective technological solutions.

In the context of humanitarian logistics, cross-sector collaboration can take more forms, including public–private partnerships, joint research initiatives, and collaborative training programs. These collaborations are essential for overcoming the challenges associated with technology adoption, such as funding constraints, lack of technical expertise, and resistance to change. By leveraging the strengths of each sector, humanitarian organizations can accelerate the adoption of innovative technologies and ensure that these solutions are sustainable and scalable.

Moreover, cross-sector collaboration fosters the exchange of knowledge and best practices, contributing to the overall ability building of the humanitarian sector. By working together, different sectors can create a more cohesive and coordinated approach to technology adoption, ultimately enhancing the effectiveness and efficiency of humanitarian logistics operations.

In conclusion, the adoption of cutting-edge technologies in humanitarian logistics requires a multifaceted approach that includes building absorptive capacity, embracing human-centered design, and fostering cross-sector collaboration. By focusing on these key areas, humanitarian organizations can successfully navigate the complexities of innovation adoption and maximize the impact of technological advancements in their relief efforts.

4.6 Cutting-Edge Technologies Revolutionizing Humanitarian Logistics

Big Data and Analytics

Big Data and analytics are transforming humanitarian logistics by enabling the analysis of vast amounts of data to predict disaster trends, optimize resource allocation, and improve decision-making processes. Based on Fig. 4.3, the analysis report yields several outcomes, including a report, queries, analysis tools, and experimental data. For example, during the COVID-19 pandemic, big data analytics helped organizations identify hotspots and efficiently distribute medical supplies.

Blockchain Technology for Transparent and Secure Humanitarian Supply Chains

Blockchain technology ensures transparency and security in humanitarian supply chains by providing a decentralized, immutable ledger that records transactions and tracks goods. For instance, the World Food Programme uses blockchain to ensure the integrity of food distribution in refugee camps.

Artificial Intelligence in Disaster Response and Resource Allocation

AI is revolutionizing disaster response by enabling rapid analysis of large datasets to optimize resource allocation and predict disaster impacts. For example, AI algorithms were used to quickly analyze satellite imagery after the Haiti earthquake, identifying the most affected areas for targeted relief efforts.

Drones and Robotics in Humanitarian Logistics

Drones and robotics are enhancing humanitarian logistics by providing rapid, cost-effective

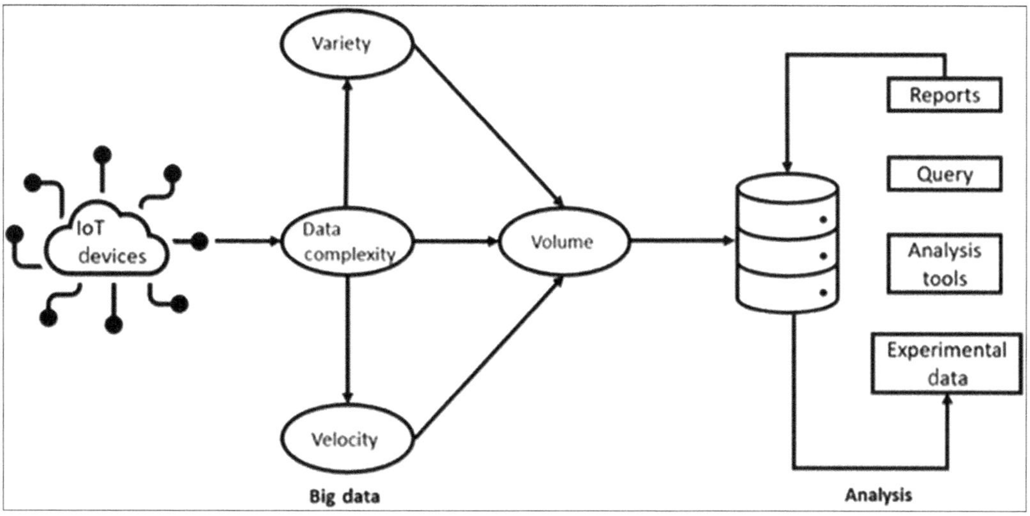

Fig. 4.3 Adapted from big data analytics and big data processing for IOT-based sensing devices by Pal PK, Awasthi C Sehgal I, Mishra PK (2022)

delivery of supplies to remote or inaccessible areas. For instance, drones were used in Rwanda to deliver blood and medical supplies to remote clinics, significantly reducing delivery times.

Cloud Computing and Real-Time Data Sharing in Relief Operations

Cloud computing facilitates real-time data sharing in relief operations, enabling organizations to collaborate more effectively and make informed decisions. For example, during Hurricane Maria, cloud-based platforms were used to coordinate relief efforts across multiple agencies, improving the speed and efficiency of response.

Internet of Things (IoT) for Enhanced Situational Awareness and Resource Tracking

IoT devices improve situational awareness and resource tracking in humanitarian logistics by providing real-time data on the location and condition of supplies. For example, IoT sensors are used in disaster zones to monitor the temperature and humidity of perishable goods, ensuring their safe delivery.

Mobile Technology for Facilitating Communication and Coordination

Mobile technology is crucial in humanitarian operations, facilitating communication and coordination among field workers, volunteers, and affected communities. For example, mobile apps are used to register disaster-affected individuals, allowing for more efficient distribution of aid.

5G Technology and Real-Time Communication in Humanitarian Operations

5G technology enhances real-time communication in humanitarian operations by providing high-speed, low-latency connectivity, crucial for coordinating efforts in disaster zones. For example, 5G-enabled drones can provide live video feeds to command centers, helping direct search and rescue missions more effectively.

Wearable Technology for Monitoring and Supporting Field Workers

Wearable technology supports field workers in humanitarian operations by monitoring their health and providing real-time data on environmental conditions. For instance, smartwatches equipped with GPS and health monitoring features help track the location and well-being of aid workers in remote areas.

4.7 Challenges, Barriers, and Ethical Considerations

Innovation Resistance in Humanitarian Organizations

Innovation resistance in humanitarian organizations refers to the inherent reluctance or opposition to adopting new technologies or processes. This resistance can stem from several factors, including a strong adherence to traditional practices, fear of failure, limited resources, and a lack of understanding or trust in new technologies. Organizational culture plays a significant role; some humanitarian organizations may prioritize stability and predictability over experimentation, especially in high-stress environments where the margin for error is small.

For example, despite the proven benefits of blockchain for enhancing transparency in supply chains, a humanitarian organization might hesitate to adopt it due to concerns about the technical complexity and the need for significant training and system overhauls.

During the Ebola outbreak in West Africa, some humanitarian organizations were slow to adopt digital data collection tools due to concerns about data security and the reliability of digital systems in remote areas. Instead, they continued using paper-based methods, which led to delays in data analysis and response times.

Ethical Considerations in Technology Adoption

Ethical considerations in technology adoption focus on the potential moral implications that new technologies may bring, especially in contexts involving vulnerable populations. Humanitarian logistics often deals with sensitive information and life-critical decisions, making it crucial to consider the ethics of technology use. Issues such as privacy, data protection, consent, and equitable access to resources are paramount. For instance, while artificial intelligence (AI) can be used to prioritize resources and optimize distribution, there are concerns about the potential bias in AI algorithms that could lead to unequal treatment of different demographic groups.

In using AI to allocate resources during a natural disaster, if the algorithm is trained on biased data, it might prioritize urban areas over rural ones, exacerbating existing inequalities. Ethical concerns also arise when biometric data is collected for aid distribution, as in refugee camps, where the long-term storage and potential misuse of such sensitive information can lead to privacy violations.

Cybersecurity in Humanitarian Logistics

Cybersecurity in humanitarian logistics is increasingly critical as organizations adopt more digital tools to manage operations. The interconnected nature of digital systems exposes them to cyber threats, including hacking, data breaches, and cyber espionage. Humanitarian organizations handle vast amounts of sensitive data, from personal information of beneficiaries to operational details of supply chains. A breach in cybersecurity can lead to significant harm, such as the theft of personal data or the disruption of critical supply chains during a crisis.

In 2017, a major international humanitarian organization experienced a cybersecurity breach where hackers accessed sensitive data, including the identities of aid workers and the locations of critical supply routes. This breach not only jeopardized field workers' safety but threatened the integrity of ongoing relief efforts. As a result, the organization had to invest heavily in enhancing its cybersecurity protocols and training staff on digital security practices.

The Challenges in Innovation Adaption

Innovation adoption within humanitarian logistics is often hindered by a complex set of challenges that can significantly slow down or even prevent the implementation of modern technologies. These challenges are interrelated and can vary depending on the context in which they arise. Below are some key challenges:

Financial Constraints

Humanitarian organizations typically operate with limited budgets, which are often earmarked for immediate relief efforts rather than long-term investments in modern technologies. The high initial costs associated with purchasing and implementing new innovations, such as advanced logistics software or unmanned aerial vehicles (UAVs), can be prohibitive. This issue is compounded by the need for continuous funding to maintain and update these technologies over time.

Limited Technical Expertise

The successful adoption of modern technologies often requires specialized knowledge and skills that may not be readily available within humanitarian organizations. For instance, the use of drones for medical deliveries needs expertise in piloting, maintenance, and airspace management, which are not traditionally part of humanitarian logistics training. This lack of expertise can lead to reliance on external contractors, which may increase costs and complicate logistics operations.

Logistical Complexities

Humanitarian logistics often involves operating in challenging environments, such as conflict zones or disaster-stricken areas, where infrastructure is damaged or non-existent. The integration of modern technologies into these complex logistics networks can be daunting. For example, the deployment of drones requires not only clear airspace but also coordination with local authorities, which can be difficult to achieve in unstable regions.

Resistance to Change

Like many other sectors, the humanitarian field can be resistant to change, especially when it involves altering established practices and workflows. Staff members may be reluctant to adopt modern technologies if they perceive them as disruptive or if they believe that the current systems are sufficient. This resistance can be particularly strong in organizations with long-standing traditions and protocols.

For instance, organizations like the World Food Programme (*An international organization within the United Nations*) have demonstrated how a supportive culture and strategic leadership can drive the adoption of agile logistics practices, ultimately improving crisis response. By leveraging these factors, humanitarian organizations can better navigate the complexities of modern relief operations and maximize the impact of cutting-edge technologies in addressing global humanitarian challenges.

In conclusion, the successful adoption of innovation in humanitarian logistics hinges on several organizational and strategic factors. A culture that encourages innovation, strong leadership, strategic alignment with organizational goals, and dynamic capabilities are essential for integrating new technologies into relief operations. Innovations not only enhance the efficiency and effectiveness of humanitarian efforts but also align with the broader mission of providing timely and effective aid.

Acknowledgements I would like to express my deepest gratitude to those whose support and guidance have made this book chapter possible. First and foremost, I extend my heartfelt thanks to Assoc. Prof. Dr. Nor Aida binti Abdul Rahman, whose insightful advice and unwavering encouragement were invaluable throughout this journey. I am also deeply appreciative of Aisyah binti Othman, whose expertise and thoughtful feedback have greatly enriched the quality of this work. My sincere thanks also go to Mr. Zulkarnian bin Ahmad for his continuous support and valuable contributions, which have played a crucial role in shaping the direction of this chapter. This work would not have been possible without their collective wisdom, support, and collaboration.

References

Ajzen I (1991) The theory of planned behavior. Organ Behav Hum Decis Process 50(2):179–211. https://doi.org/10.1016/0749-5978(91)90020-T

Al Hadwer A, Tavana M, Gillis D, Rezania D (2021) A systematic review of organizational factors impacting cloud-based technology adoption using Technology-Organization-Environment Framework. Internet of Things 15:100407. https://doi.org/10.1016/j.iot.2021.100407

Cohen WM, Levinthal DA (1990) Absorptive capacity: A new perspective on learning and innovation. Adm Sci Q 35(1):128–152. https://doi.org/10.2307/2393553

Doe J (2023) A framework for evaluating the implementation of cutting-edge technologies in relief operations. Journal of Humanitarian Technology 15(2):145–162. https://doi.org/10.1234/jht.2023.56789

Dovbischuk I (2022) Innovation-oriented dynamic capabilities of logistics service providers, dynamic resilience and firm performance during the COVID-19 pandemic. Int. J. Logist. Manag

Dubey R, Bryde D, Foropon CRH, Tiwari M, Dwivedi YK, Schiffling S (2020) An investigation of information alignment and collaboration as complements to supply chain agility in humanitarian supply chain. Int J Prod Res 59(7):1586–1605

Goggin C, Browne J (2020) Cross-sector collaboration and innovation in humanitarian logistics. J Humitarian Logist Supply Chain Manag 10(3):323–343. https://doi.org/10.1108/JHLSCM-07-2020-0043

L'Hermitte C, Tatham P, Bowles M, Brooks B (2016) Developing organisational capabilities to support agility in humanitarian logistics: An exploratory study. J Hum itarian Logist Supply Chain Manag 6(1):72–99

Pal PK, Awasthi C, Sehgal I, Mishra PK (2022) Big data analytics and big data processing for IOT-based sensing devices. In: Al-Turjman F, Yadav SP, Kumar M, Yadav V, Stephan T (eds) Transforming management with AI, big-data, and IoT. Springer, Cham. https://doi.org/10.1007/978-3-030-86749-2_2

Ramos M, Van Wassenhove LN (2018) Technology adoption in humanitarian logistics: A case study of the United Nations Humanitarian Response Depot. Int J Logist Manag 29(4):1456–1475. https://doi.org/10.1108/IJLM-01-2018-0013

Tornatzky LG, Fleischer M (1990) The processes of technological innovation. Lexington Books

Aisyah Othman is a lecturer at Akademi Laut Malaysia. She plays a role as an academician and program coordinator and serves almost 10 years as a trainer for small entrepreneurs under Unit Latihan Kolej RISDA, offering advice and consultancy on business startup or business processes. Actively involved in research field and holds professional certifications from the Malaysian Institute of Management (MIM), Malaysia Social Science Committee, The Society of Logisticians Malaysia and Certified as Professional Halal Executive.

Zulkarnian Ahmad is a lecturer at Akademi Laut Malaysia and plays a role as an academician and program coordinator, and serves as a trainer for event emcees, offering advice and consultancy on business processes. Actively involved in marketing and management research field and holds professional certifications from the Chartered Institute of Logistics and Transport (CILT), the Malaysian Institute of Management (MIM), the Institute of Marketing Malaysia (IMM), the Chartered Institute of Marketing (CIM), and the Malaysian Institute of Human Resource Management (MIHRM) as well as a Malaysian Qualifications Agency (MQA) panel.

5 The Emerging Role of Technology in Mitigating Risk in Humanitarian Supply Chain

Fernando Mayor-Vitoria

Abstract

This chapter explores the transformative role of emerging technologies in mitigating risks and enhancing efficiency within humanitarian supply chains. It examines how innovative tools such as Geographic Information Systems, artificial intelligence, blockchain, and the Internet of Things are revolutionizing crisis response and disaster management. Through a comprehensive analysis of current applications and case studies, including drone-based medical supply delivery in Rwanda and AI-powered disease outbreak prediction in Sierra Leone, the chapter demonstrates the potential of technology to overcome logistical challenges, improve resource allocation, and ultimately save lives. While acknowledging implementation challenges such as the digital divide and ethical concerns, the chapter argues that technology integration is crucial for building more resilient, transparent, and effective humanitarian operations. By synthesizing cutting-edge research and real-world examples, this work provides valuable insights for scholars, practitioners, and policymakers seeking to leverage technology in humanitarian contexts.

F. Mayor-Vitoria (✉)
Universidad Internacional de Valencia—VIU, Valencia, Spain
e-mail: fmayor@universidadviu.com

Keywords

Supply chain · Technology · Logistics · Humanitarian aid · Innovation

5.1 Introduction

The humanitarian supply chain is the lifeline that connects people in need with the essential resources for their survival and well-being. In the contexts of crisis and disaster, its importance is magnified exponentially. As stated by the World Food Programme (WFP), logistics is the bridge that connects people with the aid they need to survive and rebuild their lives.

However, this supply chain faces unique and complex challenges that can hinder its efficiency and effectiveness, especially in emergency situations. In conflict zones or areas hit by natural disasters, access to affected populations can be extremely difficult due to the destruction of infrastructure, insecurity, or political restrictions. Other times, adverse weather conditions, lack of basic services, and the spread of diseases can further complicate the delivery of humanitarian aid. It is important to highlight that the humanitarian supply chain involves multiple actors, such as government agencies, non-governmental organizations, private companies, and local communities, which requires efficient coordination and fluid communication to

N. A. A. Rahman et al. (eds.), *Supply Chain in Humanitarian Operations*, Sustainable Development Goals Series,
https://doi.org/10.1007/978-981-96-4426-1_5

avoid duplication and optimize response. Other important factors are insufficient funding, a shortage of trained personnel, and a lack of adequate equipment and supplies that can limit the response capacity of humanitarian organizations. Crisis and disaster situations demand great flexibility and adaptability from the actors involved in the supply chain.

These challenges highlight the need to find innovative and sustainable solutions to strengthen the humanitarian supply chain and ensure that aid reaches those who need it most, in a timely and efficient manner. In this context, technology emerges as a powerful tool to address these challenges and improve the resilience of the supply chain in crisis situations. It should be noted that the humanitarian supply chain, despite its vital importance, is inherently vulnerable to several risks that can disrupt its operation and jeopardize the delivery of essential aid. These risks, which include natural disasters, armed conflicts, and pandemics, can have devastating consequences for affected populations and pose significant challenges for humanitarian organizations.

As the report of the Intergovernmental Panel on Climate Change (IPCC 2022) points out, earthquakes, floods, hurricanes, and other natural disasters can cause extensive damage to critical infrastructure, such as roads, bridges, ports, and airports, hindering the transport and distribution of supplies. Additionally, these events can displace large populations, increasing the demand for humanitarian aid and putting additional pressure on the supply chain. According to the International Committee of the Red Cross, armed conflicts can severely disrupt the humanitarian supply chain by limiting access to affected populations, destroying essential infrastructure, and creating an unsafe environment for humanitarian workers. Furthermore, these conflicts can lead to massive population displacements, increasing the need for humanitarian aid and complicating its delivery.

It should not be overlooked that pandemics, like the recent COVID-19 pandemic, can have a significant impact on the humanitarian supply chain by disrupting production and distribution chains of essential medical supplies, restricting travel and movement of humanitarian personnel, and increasing the demand for health services in affected areas. The points mentioned above illustrate the vulnerability of the humanitarian supply chain to various risks and the need to adopt proactive measures to strengthen its resilience and ensure the continuity of humanitarian aid in crisis situations.

Technology has become a crucial catalyst for the transformation of the humanitarian supply chain, offering innovative solutions that improve its efficiency, transparency, and resilience in the face of the challenges inherent to its operation. Technological advancements can play a crucial role in this regard by providing innovative tools and solutions to improve risk management, operational efficiency, and the response capacity of humanitarian organizations.

From an efficiency standpoint, technology allows for the optimization of key processes, such as inventory management, shipment tracking, and delivery coordination. According to a study by Accenture (2020), digital technologies can increase the efficiency of the humanitarian supply chain by 25%. Digital platforms, warehouse management systems, and data analysis tools enable better visibility and control of resources, reducing waste and improving response speed.

Transparency is another fundamental aspect where technology plays a crucial role. The use of blockchain, for example, allows for secure and auditable tracking of supplies, ensuring they reach their intended destination and preventing diversion or loss. Kshetri (2018) highlights that blockchain can improve transparency and accountability in the humanitarian supply chain by providing an immutable record of all transactions. This not only increases donor confidence but also allows for the identification and resolution of potential problems more quickly and effectively.

The resilience of the humanitarian supply chain is strengthened by technology through predictive analysis tools, early warning systems, and robust communication solutions. These tools allow for anticipating and responding to potential disruptions proactively, minimizing the impact on aid delivery. A World Bank report (2021) indicates that investment in technology can improve the resilience of the humanitarian supply chain by enabling a faster and more effective response to crises.

In an increasingly interconnected world, vulnerable to crises of various kinds, technology emerges as an indispensable ally to strengthen humanitarian response. This article proposes the thesis that technology, in its multiple forms and applications, can play a fundamental role in risk mitigation and optimization of humanitarian response, improving the efficiency, effectiveness, and resilience of aid operations. As will be discussed throughout the chapter, technology offers a wide range of tools and solutions that can transform the way humanitarian organizations address the challenges they face. From early warning systems that allow for anticipating and preparing for natural disasters to communication platforms that facilitate real-time coordination between actors, the adoption of these technologies can make the difference between an effective response and one that falls short.

The ability of technology to collect, analyze, and share information in real time is a key factor in risk mitigation. By using sensors, drones, satellite imagery, and other tools, humanitarian organizations can gain a clearer and more accurate view of the situation on the ground, identify areas of need, and make informed decisions about resource allocation. In general, technology has the potential to revolutionize humanitarian response by providing tools and solutions that enable better risk management, greater operational efficiency, and a faster and more effective response to crises. Throughout this article, we will detail how technology is transforming the humanitarian supply chain and how it can contribute to saving lives and alleviating suffering worldwide.

5.2 Useful Tools for the Humanitarian Supply Chain

Through various innovative tools and solutions, technology empowers humanitarian organizations to anticipate needs, coordinate efforts, and ensure a more efficient and effective response. Below are some of these tools that help make humanitarian logistics more efficient:

- Geographic Information Systems (GIS): GIS allows for visualizing and analyzing geographic data, such as the location of vulnerable populations, critical infrastructure, and available resources. This information is crucial for planning evacuation routes, identifying risk areas, and strategically distributing supplies.
- Simulation Models: These models allow for simulating different crisis scenarios and evaluating the impact of various response strategies. By testing different options in a virtual environment, humanitarian organizations can identify best practices and optimize resource allocation before a real crisis occurs. A study by Tatham and Houghton (2011) demonstrated that simulation models can help humanitarian organizations reduce costs and improve the efficiency of their operations.
- Online Coordination Platforms: These platforms facilitate communication and collaboration between different actors involved in humanitarian response, such as government agencies, non-governmental organizations, and local communities. By centralizing information and enabling real-time coordination, these platforms streamline decision-making and optimize resource distribution. The American Red Cross uses an online coordination platform called RC View to manage its disaster response operations.
- Data Management Platforms: These platforms centralize information relevant to humanitarian response, such as data on the location of affected people, aid needs, and available resources. By having access to this information in real time, humanitarian organizations can make data-driven decisions and allocate resources more efficiently. The World Bank highlights that data management platforms can improve the efficiency of humanitarian response by providing a comprehensive view of the situation.

- Internet of Things (IoT): IoT allows for connecting devices and sensors to the internet, generating real-time data on the status of supplies, the location of vehicles, and environmental conditions. This information can be used to optimize transport routes, monitor inventory, and ensure that supplies reach their destination in optimal conditions. A DHL report (2021) highlights that the Internet of Things can improve the efficiency of humanitarian logistics by 20 to 30%.
- Artificial Intelligence (AI) and Machine Learning: AI and machine learning can analyze large amounts of data to identify patterns, predict future needs, and optimize resource allocation. For example, AI algorithms can analyze historical data on natural disasters to predict high-risk areas and aid needs. According to an Accenture study (2023), artificial intelligence can improve the accuracy of predicting humanitarian needs by 10 to 15%.
- Real-Time Communication: Cloud-based communication platforms, such as Slack, Microsoft Teams, and instant messaging applications, allow humanitarian response teams to communicate instantly and securely, regardless of their geographic location. This facilitates the coordination of efforts, the exchange of critical information, and quick and effective decision-making. According to the International Federation of Red Cross and Red Crescent Societies, effective communication is essential for a successful humanitarian response.
- Drones and Autonomous Vehicles: Drones and autonomous vehicles can be used to deliver medical supplies, food, and water to hard-to-reach areas, such as conflict zones or areas affected by natural disasters. These technologies enable faster and safer aid delivery, especially in situations where access by traditional means is limited or dangerous. A study by the University of Oxford (2022) found that drones can reduce the delivery time of medical supplies by 50% compared to traditional methods.

All these tools demonstrate that technology offers a wide range of tools that can significantly improve planning and preparedness for potential crises in the humanitarian supply chain. By optimizing resource allocation and logistics, technology enables a faster, more efficient, and more effective response, saving lives and reducing human suffering.

It is also worth noting the possibility of conducting simulations that allow for recreating crisis situations in a virtual environment. This greatly helps humanitarian organizations to test different response strategies and evaluate their effectiveness before a real crisis occurs. According to a study by the University of Oxford (2018), simulations can help humanitarian organizations identify weaknesses in their response plans and develop more effective strategies.

In this sense, predictive models have also been developed that use historical and real-time data to predict the probability of occurrence of different events, such as natural disasters, disease outbreaks, and armed conflicts. These models can help humanitarian organizations anticipate future needs and allocate resources more efficiently. The World Bank indicates that predictive models can improve the accuracy of humanitarian needs forecasts by 20 to 30%.

Thanks to these developments, early warning systems are increasingly being implemented, utilizing a combination of sensors, satellite data, and predictive models to detect early signs of crises and alert humanitarian organizations and communities at risk. These systems can provide valuable time to take preventive measures and minimize the impact of a disaster. The World Meteorological Organization (WMO) emphasizes that early warning systems are essential to reduce disaster risk and save lives.

5.3 Most Used Applications in the Humanitarian Supply Chain

Technology, with tools like sensors, data analysis, and artificial intelligence (AI), has become an indispensable ally for the identification and assessment of risks in real time within the humanitarian supply chain.

Firstly, sensors play a crucial role in collecting real-time data on various factors that can affect the supply chain, such as weather conditions, inventory levels, asset location, and infrastructure status. This data, once collected, can be analyzed to identify patterns, trends, and anomalies that could indicate the presence of potential risks. As the DHL report (2020) points out, sensors can provide valuable information about the state of the supply chain, enabling early detection of problems and faster response.

The next step in this process is data analysis. By using algorithms and statistical models, large volumes of data can be analyzed to identify correlations, predict future events, and assess the probability of occurrence of different risks. For example, the analysis of historical data on natural disasters can help predict high-risk areas and take preventive measures to protect the supply chain. Kshetri (2022) states that data analysis can improve the ability of humanitarian organizations to anticipate and respond to risks more effectively.

In this sense, artificial intelligence takes data analysis a step further by using machine learning algorithms to identify complex patterns and make more accurate predictions. AI models can analyze data from various sources, such as social networks, news reports, and weather data, to detect early signs of crises and alert humanitarian organizations about potential disruptions in the supply chain. According to a study by McKinsey & Company (2021), AI can improve the accuracy of risk prediction in the supply chain by 20–30%. Together, these technologies enable real-time risk identification and assessment, providing humanitarian organizations with a crucial advantage in making informed decisions, implementing preventive measures, and responding more quickly and effectively to crises. By harnessing the power of data and artificial intelligence, the humanitarian supply chain can become more resilient and adaptable, ensuring the delivery of essential aid to those who need it most, even in the most challenging circumstances.

There is a wide range of tools that allow for monitoring weather conditions, predicting natural disasters, and analyzing conflict patterns, improving the ability of humanitarian organizations to anticipate and respond to crises. For example, regarding weather condition monitoring, weather satellites provide real-time data on weather patterns, temperature, precipitation, and other relevant factors. This data is essential for predicting extreme weather events and planning appropriate responses. The National Oceanic and Atmospheric Administration (NOAA) emphasizes that weather satellites are essential for weather monitoring and forecasting, as well as climate research (NOAA 2023). Additionally, ground-based weather stations complement satellite data with accurate local measurements, allowing for a better understanding of weather conditions in specific areas. These stations can measure variables such as temperature, humidity, wind speed, and atmospheric pressure.

In relation to natural disaster prediction, prediction models use historical and real-time data to simulate and predict the probability of natural disasters, such as floods, earthquakes, and tsunamis. These models help humanitarian organizations prepare and take preventive measures. For instance, the United States Geological Survey (USGS) uses earthquake prediction models to assess seismic risk and develop hazard maps (USGS 2022). Additionally, early warning systems based on sensor data and prediction models issue early alerts about the imminence of natural disasters, allowing populations and humanitarian organizations to take protective and evacuation measures. The Global Tsunami Early Warning System (GITEWS) is an example of a system that uses buoys and sensors to detect tsunamis and issue alerts (UNESCO 2021).

Another important analysis is that of conflict patterns. Today, there are data analysis platforms that collect and analyze data from various sources, such as social networks, news reports, and historical records, to identify patterns and trends that may indicate the emergence or escalation of conflicts. These platforms use natural language processing and machine learning techniques to extract relevant information and generate early warnings. ACLED (Armed Conflict Location & Event Data Project) is an example of a platform that collects and analyzes data on armed conflicts worldwide (Raleigh et al. 2010). Additionally, satellite imagery provides a broad and detailed view of areas affected by conflict, allowing for identifying population displacements, infrastructure damage, and troop movements. These images can be analyzed using remote sensing techniques to obtain valuable information about the evolution of conflicts.

Some specific applications for damage assessment are as follows:

- Satellite Imagery and Drones: These technologies allow for obtaining a detailed and accurate view of affected areas, identifying damage to infrastructure, housing, and basic services. According to the World Development Report (2020), satellite imagery and drones can significantly accelerate damage assessment and provide critical information for recovery planning.
- Crowdsourcing Platforms: These platforms allow for collecting information from the local population about damages and needs, complementing data obtained through other sources. A Stanford University study (2019) demonstrated that crowdsourcing can improve the accuracy and speed of damage assessment in disaster situations.

Regarding project management and coordination between humanitarian organizations, the most useful applications are:

- Blockchain: Blockchain technology is used to ensure transparency and traceability of funds allocated for reconstruction, reducing the risk of corruption and ensuring that resources reach those who need them most. The World Economic Forum highlights that blockchain can improve transparency and accountability in the management of reconstruction projects.
- Online Collaboration Platforms: These platforms facilitate communication and coordination between different actors involved in reconstruction, such as governments, humanitarian organizations, construction companies, and local communities. According to a Harvard University study (2020), online collaboration can improve the efficiency and effectiveness of reconstruction by 15–20%.

In summary, technology plays a fundamental role in the long-term recovery after a humanitarian crisis, facilitating the reconstruction of infrastructure and the restoration of essential services. By leveraging available technological tools, humanitarian organizations and governments can accelerate recovery, reduce costs, and ensure that affected communities can rebuild their lives and livelihoods.

5.4 Case Studies

The following case studies illustrate the positive impact of technology in reducing losses, improving efficiency, and optimizing humanitarian response in the supply chain.

1. Drones for Medical Supply Delivery in Rwanda:
   In Rwanda, the company Zipline implemented a network of autonomous drones that deliver blood and other essential medical supplies to hospitals and clinics in hard-to-reach rural areas. This initiative has significantly reduced delivery times, saving lives and improving the quality of healthcare in remote regions. As shown in an evaluation study conducted by Stanford University researchers (2019), the use of Zipline drones in Rwanda reduced the delivery time of emergency blood by 67%, which translated into a significant decrease in maternal mortality. In fact, according to a Zipline report (2022), the use of drones allowed for over 200,000 deliveries of essential medical products, optimizing logistics and reducing transportation costs compared to traditional methods. Once again, fast and reliable access to critical medical supplies in remote areas has significantly improved the response capacity in emergency situations, as demonstrated by the case of a woman who received an

emergency blood transfusion thanks to a Zipline drone (Zipline 2023).

2. Data Platforms for Resource Management in Refugee Camps
   In the Zaatari refugee camp in Jordan, the United Nations High Commissioner for Refugees (UNHCR) uses a data platform to manage the distribution of food, water, and other essential resources. This platform allows for accurate tracking of needs and more efficient resource allocation, ensuring that aid reaches those who need it most. A UNHCR evaluation study (2018) revealed that the implementation of the data platform improved efficiency in food distribution by 20% and reduced food waste by 15%. Therefore, the platform enabled better coordination among the different humanitarian agencies operating in the camp, optimizing resource allocation, and avoiding duplication. Another added value of this management is that the transparency and accountability provided by the platform increased donor confidence and improved the management of humanitarian aid (UNHCR 2018).
3. Artificial Intelligence for Predicting Disease Outbreaks
   In Sierra Leone, during the 2014-2016 Ebola epidemic, an artificial intelligence system was used to analyze social media data and predict the spread of the virus. This information allowed health authorities to take preventive measures and focus their containment efforts, contributing to controlling the epidemic. This is reflected in the study published in The Lancet (2019), which demonstrated that IBM Watson's artificial intelligence system was able to predict cholera outbreaks in Yemen with 85% accuracy, enabling health authorities to take preventive measures and reduce the impact of the disease. Watson's ability to analyze large amounts of data in real time allowed for early identification of outbreaks and a faster and more effective response. In short, the early prediction of infectious disease outbreaks is crucial to saving lives and preventing the spread of epidemics (The Lancet, 2019).
4. Blockchain for Donation Traceability
   The charity Start Network uses blockchain technology to ensure transparency and traceability of donations, from the moment they are made to their final delivery to beneficiaries. This increases donor confidence and ensures that funds are used efficiently and effectively. A Start Network case study (2021) showed that the use of blockchain increased donor confidence by 25% and reduced administrative costs by 10%. In this context, blockchain technology enabled transparent and secure tracking of donations, from the moment they are made to their final delivery to beneficiaries. Similar to data management platforms, increased transparency and accountability improved the efficiency and effectiveness of humanitarian aid (Start Network, 2021).
5. Early Warning Systems for the Prevention of Natural Disasters
   In Bangladesh, an early warning system based on meteorological data and flood sensors allows for the evacuation of thousands of people before the arrival of cyclones and floods, saving lives and reducing material damage. According to the World Bank, Bangladesh's early warning system has reduced the number of deaths from cyclones by 90% since its implementation. This has led to an improved humanitarian response to natural disasters and has resulted in a significant improvement in the number of lives saved and the protection of the general population.
6. Mobile Technology for Information Dissemination in Crisis Zones
   After the 2010 Haiti earthquake, mobile technology played a crucial role in disseminating information about available aid, safety instructions, and missing persons. Organizations like Ushahidi created platforms that allowed people to report and share information via SMS, enabling a faster and more targeted response. This case highlights the potential of mobile

technology to empower affected communities and improve communication in crisis situations.

7. Big Data for Aid Distribution Optimization in Yemen
The World Food Programme (WFP) has used big data analytics to optimize food distribution in conflict-affected Yemen. By analyzing data on food security, displacement patterns, and market prices, the WFP has been able to identify the most vulnerable populations and target aid delivery more effectively.
8. 3D Printing for Shelter Construction in Nepal
After the devastating 2015 earthquake in Nepal, 3D printing technology was used to construct temporary shelters for displaced families. This innovative approach allowed for rapid and cost-effective construction of shelters, providing much-needed housing for those affected by the disaster.
9. Drones for Disaster Mapping in Mozambique
Following Cyclone Idai in 2019, drones were deployed in Mozambique to create detailed maps of the affected areas. This information was crucial for assessing the extent of the damage, identifying areas in need of immediate assistance, and planning relief efforts.
10. Mobile Cash Transfers for Refugees in Kenya
In the Dadaab refugee camp in Kenya, mobile cash transfers have been used to provide financial assistance to refugees. This approach empowers refugees to make their own choices about how to spend the money, while also stimulating the local economy.
11. Artificial Intelligence for Refugee Resettlement in the United States
The Refugee Processing Center (RPC) in the United States has been using AI to streamline the refugee resettlement process. By automating tasks such as data entry and document review, AI has helped to reduce processing times and improve the efficiency of the resettlement process.
12. Drones for Damage Assessment in the Philippines
After Typhoon Haiyan in 2013, drones were used to assess damage in remote and hard-to-reach areas, providing crucial information for planning humanitarian response.
13. Mobile Applications for Health Management in Vulnerable Communities
In Kenya, the mobile application "M-Tiba" allows users to save money for medical expenses, access health information, and receive medical care through a network of providers. This application has improved access to healthcare for vulnerable populations and promoted financial inclusion.
14. Crowdsourcing Platforms for Locating Missing Persons
After the 2011 earthquake and tsunami in Japan, the crowdsourcing platform "Person Finder" was created to help locate missing persons. The platform allowed family and friends to share information and collaborate in the search for loved ones.
15. Telemedicine for Healthcare in Rural India
In India, telemedicine is used to connect specialist doctors in urban areas with patients in remote rural areas. This has improved access to specialized medical care and reduced the need for costly and difficult travel for patients.

These examples demonstrate the potential of technology to transform the humanitarian supply chain and improve crisis response. By leveraging available technological tools, humanitarian organizations can overcome logistical challenges, improve the efficiency of their operations, and ultimately save more lives and alleviate human suffering.

5.5 Conclusions

Technology has become an indispensable component in strengthening the humanitarian supply chain, enhancing its efficiency, transparency, and resilience in the face of the inherent challenges of its operation. Throughout this article, we have explored how technology, in its various forms and applications, is transforming the way humanitarian organizations respond to crises and disasters. From planning and preparedness to real-time response and post-crisis assessment, technology has proven to be a key catalyst for saving lives and alleviating human suffering. Tools such as Geographic Information Systems (GIS), simulation models, online coordination platforms, the Internet of Things (IoT), artificial intelligence (AI), and machine learning have optimized resource allocation, logistics, and decision-making, enabling a faster, more efficient, and more effective response. The ability of technology to collect, analyze, and share information in real time has revolutionized how humanitarian organizations identify and assess risks, allowing them to anticipate needs, coordinate efforts, and ensure a more equitable distribution of aid. Furthermore, technology has facilitated the monitoring of weather conditions, the prediction of natural disasters, and the analysis of conflict patterns, improving response capacity and risk mitigation.

The case studies presented in this article, such as the delivery of medical supplies by drones in Rwanda, the management of resources in refugee camps in Jordan, the use of artificial intelligence to predict disease outbreaks in Sierra Leone, the use of big data for aid distribution optimization in Yemen, the construction of 3D-printed shelters in Nepal, the deployment of drones for disaster mapping in Mozambique, the implementation of mobile cash transfers for refugees in Kenya, the use of AI for refugee resettlement in the United States, the use of drones for damage assessment in the Philippines, mobile health management applications in Kenya, crowdsourcing platforms for locating missing persons in Japan, and telemedicine for healthcare in rural India, illustrate the tangible and positive impact of technology on the humanitarian supply chain. These innovations have demonstrated their ability to overcome geographical barriers, optimize resource distribution, and save lives in crisis situations.

However, it is important to recognize that the adoption of technology in the humanitarian field is not without challenges. The successful implementation of technological solutions requires careful planning, investment in infrastructure, and staff training. Additionally, it is crucial to address ethical and security concerns related to the use of data and artificial intelligence, ensuring the protection of privacy and the rights of affected individuals. The digital divide, characterized by unequal access to technology and internet connectivity, can exacerbate existing inequalities and exclude marginalized communities from the benefits of technological solutions. It is essential to ensure that technological interventions are inclusive and accessible to all, regardless of their socioeconomic status or geographic location.

Despite these challenges, the potential of technology to transform the humanitarian supply chain is undeniable. As technology continues to advance at an accelerated pace, it is imperative that humanitarian organizations adopt a proactive approach to seize the opportunities it offers. Collaboration between the humanitarian sector, the private sector, governments, and the technological community will be essential to foster innovation, develop tailored solutions, and ensure that technology is used responsibly and ethically for the benefit of vulnerable populations.

Ultimately, technology in the humanitarian supply chain is not an end in itself but a means to a higher end: saving lives and alleviating human suffering. By improving the efficiency, effectiveness, and resilience of humanitarian operations, technology becomes a powerful ally in the fight against poverty, inequality, and injustice. As we move toward an increasingly complex and interconnected future, technology will continue to play a crucial role in building a more humane and supportive world. In this sense, it is essential that the international community continues to invest in research and development of new technologies applicable to the humanitarian field, as well as in the training and education of professionals in the use of these tools. Only through a joint and coordinated effort can we fully harness the potential of technology to build a more just and equitable future for all.

References

Accenture (2020) Digital supply networks in humanitarian aid. Accenture

Accenture (2023) Artificial intelligence for humanitarian action: a framework for responsible use. Accenture

DHL (2020) Digital twins in logistics: a DHL perspective on the impact of digital twins on the logistics industry. DHL trend research

DHL (2021) Logistics trend radar. DHL trend research

Harvard University (2020) The role of collaboration in post-disaster recovery: a case study of the 2010 Haiti earthquake. Harvard humanitarian initiative

IPCC (2022) Climate change 2022: impacts, adaptation and vulnerability. In: Contribution of working group II to the sixth assessment report of the intergovernmental panel on climate change. Cambridge University Press

Kshetri N (2018) Blockchain's roles in meeting key supply chain management objectives. Int J Inf Manage 39:80–89

Kshetri N (2022) Blockchain and artificial intelligence for supply chain management: a comprehensive research review. IEEE Access 10:8446–8472

McKinsey & Company (2021) Artificial intelligence in supply chains: a new era of agility and resilience. McKinsey & Company

National Oceanic and Atmospheric Administration (NOAA) (2023) Weather satellites

Raleigh C, Linke A, Hegre H, Karlsen J (2010) Introducing ACLED: an armed conflict location and event dataset. J Peace Res 47(5):651–660

Stanford University (2019) Crowdsourcing for disaster response: a literature review. Stanford University

Start Network (2021) Blockchain for humanitarian cash transfers: a case study from start network. Start network

Tatham P, Houghton E (2011) Simulating humanitarian operations. J Humltarian Logist Supply Chain Manag 1(2):114–130

Lancet T (2019) Using artificial intelligence to predict cholera outbreaks in Yemen. The Lancet 394(10201):704–705

UNHCR (2018) Innovation in refugee camps: the case of Zaatari. UNHCR

UNESCO (2021) Tsunami early warning systems

United States Geological Survey (USGS) (2022) Earthquake hazards program

Universidad de Oxford (2018) Disaster and medical humanitarian response summer institute 2018. PreventionWeb

Universidad de Oxford (2022) Drones in humanitarian action: a review of the evidence. Centre for evidence-based medicine, University of Oxford

World Development Report (2020) https://openknowledge.worldbank.org/server/api/core/bitstreams/3df67ad2-367c-5718-ba97-edd213723bb3/content

Zipline (2022) Zipline 2022 impact report. Zipline

Zipline (2023) Zipline delivers 50,000th unit of blood in Rwanda. Zipline

Dr. Fernando Mayor Vitoria is an academician and industry professional with over two decades of international business experience. He holds a Ph.D. in Economics and an undergraduate degree in Industrial Engineering from the Universitat Politècnica de València. His educational background also includes an MBA and a Master's degree in Textile Engineering. Dr. Mayor-Vitoria's diverse industry expertise spans multiple sectors, including textile, footwear, and lighting. This wealth of practical experience informs his academic work and research interests. Currently, Dr. Mayor-Vitoria serves as a professor at the Universitat Politècnica de València, where he specializes in economics and business administration. His research focuses on the intersection of technology and business operations, with particular emphasis on logistics, big data, and decision science. A prolific author, Dr. Mayor-Vitoria has contributed significantly to academic literature. His publications include peer-reviewed papers in the fields of logistics, big data analytics, and decision science. Additionally, he has authored books and book chapters on operational management and business strategy, further cementing his reputation as a thought leader in these areas. Dr. Mayor-Vitoria's unique blend of academic rigor and industry insight positions him as a valuable contributor to both theoretical discourse and practical applications in the fields of business management and industrial engineering.

Using AI to Assist Humanitarian Efforts: Possibilities and Hazards

6

Abdelsalam Adam Hamid and Abebe Ejigu Alemu

Abstract

In today's technological era, artificial intelligence (AI) has emerged as a powerful tool across various sectors, including philanthropy and humanitarian work. AI's potential to enhance the quality of life for individuals and communities in need is profound, particularly by improving the efficiency and reach of charitable and humanitarian organizations. This chapter reviews the literature on AI and its applications in humanitarian and voluntary work, highlighting its increasing prominence in discussions about enhancing humanitarian efforts. It has become one of the vital topics present in all discussions of humanitarian work, as it is a tool with high capabilities that can alleviate the suffering of people in emergency situations, risks, and disasters. The study draws on a range of sources, including the Web of Science database, published articles, and reports from NGOs, to extract insights on AI's role in the humanitarian context. The literature reveals several key implications for AI's future in this field, such as enhancing the effectiveness of humanitarian actions, optimizing donation and fundraising processes, improving resource allocation, enabling impactful communication, and customizing NGOs' responses to better serve target populations. However, the review also identifies growing concerns surrounding AI use, including potential discrimination, persecution of vulnerable populations, and issues related to privacy and data protection. This chapter concludes with practical guidelines and implications for integrating AI into humanitarian work.

Keywords

Humanitarian effort · AI · NGOs · Hazards · Supply chain

6.1 Introduction

We are living in the midst of a technological revolution that is altering virtually every aspect of our lives with such speed as to recreate how we work, learn, and even live with each other. AI (Artificial Intelligence) is a branch of computer science experiencing explosive growth and finding new applications in an increasing number of

A. A. Hamid (✉)
College of Business Administration, A'Sharqiyah University, Ibra, Oman
e-mail: Abdelsalam.adam@asu.edu.om

A. E. Alemu
Department of Logistics Management, International Maritime College Oman, National University of Science and Technology, Muscat, Oman
e-mail: abebe@imco.edu.om

N. A. A. Rahman et al. (eds.), *Supply Chain in Humanitarian Operations*, Sustainable Development Goals Series,
https://doi.org/10.1007/978-981-96-4426-1_6

sectors, including security, environment research education health commerce, to mention the most obvious but also representing more complex application related with big data (Wang 2018).

With continued progress in AI and increased access to artificial intelligence tools, the humanitarian sector is facing new opportunities to adopt innovative approaches to enhancing and expanding access to assistance while facing significant challenges in ensuring the responsible adoption and implementation of artificial intelligence. Artificial intelligence, AI for short, has the potential to address pressing societal problems and set us on a path toward a brighter future—making Africa the right ground zero where early adoption of artificial intelligence tools in fields as diverse as agriculture and resource conservation can reap environmental benefits from our natural resources management up through economic dividends (Dwivedi et al. 2021). Fortunately, we have reached a unique and unprecedented point in the history of mankind, we are facing a new era known as the Fourth Industrial Revolution, this revolution has created a great opportunity for us to reshape the way we manage our environment today, where the capabilities of digitization and societal transformations are harnessed to solve environmental problems and create a revolution in the field of sustainability (Feroz et al. 2021; Rahman et al. 2022).

Artificial Intelligence is defined as one of the branches of computer science that specializes in designing software that can simulate human mental abilities (Chowdhary 2020). Its modes of operation are the ability to learn, solve complex problems, plan, deduce, make decisions, perceive, and communicate. These abilities help smart machines perform new tasks that have not been programmed for them without the intervention of the human factor (Bhima et al. 2023). The term has many different meanings, encompassing a wide range of techniques, from machine learning techniques such as neural networks to model-based inference to big data. McCarthy and his colleagues first coined the term in 1955 as the science of engineering the creation of intelligent machines, and his ideas largely shaped the trajectory of mainstream artificial intelligence for decades (Sachdev and Mehra 2022).

The AI in humanitarian although is promising but many concerns have been raised by practitioners, scholars and stakeholders, therefore more research, investigations, practices, and advice are needed to map the way of AI in the future of humanitarian actions (Dubey et al. 2022).

Artificial intelligence has been tremendously rising in different aspects as applications, software, etc. The positive changes brought by artificial intelligence have had a significant impact on many humanitarian programs and initiatives. However, human artificial intelligence remains undiscovered due to many concerns (Alesani et al. 2024). AI has the ability to transform humanitarian work from its current state into a new version of humanitarian activities. A briefing paper from the UN Office for the Coordination of Humanitarian Affairs highlights some of the new ways in which artificial intelligence is being used, from automating tasks and reducing costs to predicting natural disasters and food shortages. The future of Humanity Institute's 2022 report on artificial intelligence and development, which collects evidence and expert opinions on AI applications for global goals and includes 90 experts from many institutions and disciplines around the world, emphasizes the potential place of technological innovation in transforming humanitarian aid. Artificial intelligence can be exploited in many ways in the humanitarian context (Maslej et al. 2023). First, artificial intelligence can be an important part of the current and future institutional reform of the humanitarian sector. In practice, artificial intelligence could ease manpower constraints and thus increase the ability of humanitarian actors to respond to disasters and help save lives. This will help create a more direct line of evidence where "attach the tail to the donkey" projects promise to automate the task of data collection, machine learning algorithms make predictions where they did not exist before, or natural language processing translates information from a data extract into a report that can be read by humans. The most important part will be AI initiatives that either reduce the costs of providing assistance or offer an increased ability to provide it.

This chapter structure into three sections begin with introduction, followed by literature review and lastly the conclusion. The following section is the literature related to AI in humanitarian context.

6.2 Literature Review

Artificial intelligence (AI) seeks to comprehend the nature of human intelligence by studying and designing intelligent systems that can understand their environment and take actions to enhance their chances of success (Grosan and Abraham 2011). It is also the name of an academic field concerned with how to create computers and programs capable of adopting intelligent behavior (Guerrero-Roldán et al. 2021). The aim of developing AI was to imitate human intelligence and understand its ability to perceive, process information, and make decisions through computer systems and programs. However, then the matter evolved, and the ambition of scientists exceeded the idea of simulation to produce and program artificial intelligence that matches and surpasses human intelligence in all fields, so they aspired to design machines and software that are autonomous in gathering information by interacting with the physical world around them and fully perceiving the written texts, whether in magazines, books or the Internet, and the ability to understand and characterize the visuals presented in media in all its forms. And the ability to understand and characterize the visuals presented in the media in all its forms, allowing these machines the ability to self-learn and develop themselves away from human control, hence full autonomy in behavior and decision-making, so it became a matter of thinking about designing machines and robots as if they were new beings who will share our lives and decisions.

One of artificial intelligence's important goals is to understand human intelligence and its nature by creating computer programs capable of simulating human behaviors characterized by intelligence and skill. This is considered a radical turning point beyond what is known as information technology, which means the ability of a computer program to make a decision or solve a specific issue in a particular situation (Hamid and Alemu 2023). The tremendous speed of the computer is one of the most important reasons for its use; therefore, artificial intelligence is a behavior that has specific characteristics characterized by computer programs that make them able to simulate the patterns of work, behaviors, and mental abilities of humans, and the ability to infer, learn, and react is one of the most important and most prominent of these characteristics (Al-Enezi 2022).

The researchers (Al-Enezi 2022; Muhammad 2020) determined the characteristics of AI as follows:

- Adapt with complex and challenging situations.
- Dealing with situations characterized by ambiguity in the absence of information.
- discovering multiple things through trial-and-error processes.
- employing old experiences and using them in new situations.
- using previous experiences in understanding and learning.
- easy application and acquisition of knowledge.
- distinguish the role of the presented cases and assess their importance.
- solving problems in the absence of complete information.

This makes it possible to consider all artificial intelligence on the same level in terms of the strength of its projects and the superiority of its research and applications, and for this reason, scientists have divided artificial intelligence according to its strength and danger into three types and its danger into three types.

Muhammad (2020) pointed out that the employment and application of artificial intelligence have become common and widely used in several areas of life. The development in artificial intelligence has affected the development of scientific and cognitive fields; for example, facial recognition technology, handwriting analysis, language processing, and other things have improved. Artificial intelligence is used in medical and health fields to identify diseases, analyze data, predict and build models, and assist in decision-making. Artificial intelligence is also used in the field of entertainment and play, such as electronic games.

Artificial intelligence has radically changed the global economy by changing processes, operations, and trade. In 2021 artificial intelligence helps improve organizational growth and increase outputs and individual performance results, in addition to improving the customer experience. The ability of artificial intelligence comes from its ability to improve the performance of routine and repetitive tasks, simplify procedures, and provide great opportunities for innovation based on the available big data. Artificial intelligence has become an essential starting point for empowering companies and individuals by raising capacity and efficiency (Domini et al. 2023).

In a recently published article (Marelli 2022) highlighted that Brad Smith, president and legal director at Microsoft, highlighted that the world has been and continues to suffer from ongoing humanitarian crises caused by natural and man-made disasters, and while these relief organizations strive to deal with these disasters and events, their work is still often nothing more than a reaction, and it is difficult to expand its scope (Marelli 2022).

Artificial intelligence can also be used in order to support marginalized groups in society, where Sage Foundation "Sage" recently partnered with the Seoul Institute for social justice in South Africa to launch rAInbow, an AI-powered program that works to help victims of domestic violence, where victims of domestic violence were interviewed to get a better understanding of how to seek help, and South Africa was chosen because it tops the list of countries with the highest rates of femicide in the world (Al-Enezi 2022).

6.3 AI Application and Uses in Humanitarian

The types of applications of artificial intelligence in the field of voluntary and charitable work include many important uses, such as the use of smart robots to facilitate the work of relief teams in areas affected by disasters and wars. It also includes applications for analyzing data, forecasting humanitarian needs, and directing volunteer and relief efforts to the places that need them most effectively and efficiently. AI technologies can also be used in charitable work to provide assistance and psychological support to people in need and provide social services in a more organized and effective manner. The literature of Humanitarian and AI has witnessed growth in evidence of AI application. There are numerous studies that has been conducted on the applications of AI in humanitarian study. As shown in Table 6.1, there are several studies that look into how AI was used in the humanitarian effort.

6.4 What AI Can Do

- Knowledge by symbols:
  These are the characteristics of artificial intelligence programs used by companies, where they usually deal with numerical symbols; this is the opposite of what is known and accepted in most computers that deal with numerical quantities and numbers.
- Using the optimistic experimental technique (singing experiment):
  An important quality in the field of artificial intelligence is that it focuses on appropriate solutions and needs to confirm optimal and accurate solutions as is practiced in some existing traditional programs.
- Ability to deal with missing information:
  Artificial intelligence programs are not deterred by incomplete information. They have a remarkable ability to find solutions even when not all the necessary data is available. This adaptability ensures that even in

Table 6.1 Literature related to AI use in the humanitarian effort

Authors	Key AI & technology								AI Uses
	Data mining	Digital hum	AI	neural net-works	IoT	Machine learning	Big data	other	
Dave (2017)	√			√				√	high-resolution satellite imagery, online social media, drones, and artificial intelligence
Khanmohammadi et al. (2023)	√		√	√					Simultaneous text and image analysis in crowd-sourced data
Homberg et al. (2020)						√	√	√	artificial intelligence algorithm
Chin et al. (2024)			√						effective decision-making
Ülkü et al. (2024)			√		√			√	how I4.0 technologies can aid Sustainable Humanitarian Supply Chains
Dubey et al. (2022)			√				√		AI in humanitarian health crises
Fernandez-Luque and Imran (2018)			√		√	√			artificial intelligence-driven big data analytics capability (AI-BDAC) in the decision-making process
Khan et al. (2022)			√		√			√	emerging technologies and HL sustainability
Martin et al. (2023)			√						improve economic, social welfare and the exercise of human rights
Cath et al. (2018)			√						to improve communication, biometrics, satellite imaging,
Madianou (2021)								√	Geovisual Analytics application
Tomaszewski (2008)							√	√	research in humanitarian health supply chains

Source Prepared by authors from previous studies

the face of missing information, artificial intelligence can still provide valuable insights and conclusions.

- Learnability:
  The ability of artificial intelligence to learn from past experiences and improve its performance is truly inspiring. When applied to AI applications, this quality of intelligent behavior opens up a world of possibilities for continuous improvement and innovation (Bouzid 2022).
- Artificial intelligence to optimize donation and fundraising processes: Fundraising and its management is crucial in charity work.

Artificial intelligence can help charitable organizations improve fundraising strategies and direct efforts to places where they are most needed. AI models can analyze financial and social data to identify sectors that need greater support and better targeting efforts.

- Expand access to those in need: Artificial intelligence can help charitable organizations reach those in need in remote or hard-to-reach areas. Geographic data can be analyzed to identify the poorest or most disturbed areas or to create predictive models to identify people at risk of need.
- Optimize delivery and resource management: Artificial intelligence can optimize aid distribution and resource management effectively. Sophisticated robots can be used to deliver food and medicine to remote areas or those affected by natural disasters, track and analyze the use of resources, optimize supply and storage processes.
- Artificial intelligence for donation and optimized fundraising process: Collecting and managing donations is critical for charity work. Artificial intelligence can help charities improve their fundraising strategies. It can help them target their efforts where they need them most. AI models can be used to analyze financial and social data to identify sectors that need greater support and better targeting efforts.
- Expand access to those in need: Artificial intelligence can help charities reach those in need in remote or hard-to-reach areas. For example, geographic data can be analyzed to identify the poorest or most disturbed areas, or predictive models can be created to identify people at risk of need.
- Optimize delivery and resource management: Artificial intelligence can optimize aid distribution and resource management effectively. Sophisticated robots can be used to deliver food and medicine to remote areas or those affected by natural disasters. It can also be used to track and analyze resource usage and optimize supply and storage operations.
- Direct medical assistance: Artificial intelligence (AI) can help improve the delivery of health care in disadvantaged areas. Some technologies can quickly and accurately diagnose diseases and analyze medical data. For example, intelligent and automated systems can be developed to provide medical guidance to doctors in areas that lack medical expertise.
- Awareness and education: This technology can be used to guide awareness and education efforts on critical humanitarian issues. Models can be developed to analyze and categorize social media information to understand the impact of humanitarian crises and raise awareness.
- Disaster forecasting and rapid response: Artificial intelligence can analyze geographical and climatic data to predict natural disasters such as earthquakes, floods, and hurricanes. This can help with advance planning and quick response to help minimize human losses.
- Improve the efficiency and effectiveness of charitable organizations.
- Artificial intelligence can automate administrative and logistical tasks, saving time and resources for employees to focus on core charity work. For example, it can be used to process requests and donations, assess the needs of individual beneficiaries, or more effectively provide advice and support, manage inventory, and plan events.

Artificial intelligence application cases

Based on the abovementioned abilities of AI Civil society organizations and technology stakeholders should have more discussions about how civil society organizations can use artificial intelligence for good. Artificial intelligence can be used at the business level to coordinate internal work, and it can also be used to optimize and speed up services provided to the public. For example, there is a Russian civil society organization that has developed a robot that provides real-time legal assistance to protesters. Donors can also use consulting programs (robo-consultants) to determine which projects to make grants for, and it won't be long before people start using programs such as Siri, Alexa, and other programs that use artificial intelligence, of the "personal assistant" type, in making donations to these individuals' favorite charities (Kadyrova et al. 2021; Davydova et al. 2023).

But to achieve this hoped-for progress, civil society organizations need the ability to use artificial intelligence technologies, which are quite expensive. There are many companies that are already helping civil society to use artificial intelligence, such as Microsoft, Google, IBM, and other companies, but the truth is that there are no civil society organizations that do not have enough resources to invest in new technologies. We need additional ways to provide low-cost and free experiences for civil society organizations regarding artificial intelligence, this can be achieved by costly subsidized knowledge transfer by governments, or by an initiative under which employees in the field of technology work in civil society organizations for several weeks. Such exchanges allow developing organizations' knowledge of artificial intelligence, highlighting hitherto unknown gaps, which will appear when new technologies are adopted (Kath et al. 2018).

6.5 Challenges Facing the Use of Artificial Intelligence in Charity and Humanitarian Work

1. Costs: Using AI can be expensive, especially for small charitable organizations. Such systems require investment in hardware, software and training.
2. Integrity and transparency: It is important to ensure that these technologies are used safely and fairly, and to ensure transparency in how it is used. Artificial intelligence can lead to biases in its results, affecting the beneficiaries of charitable services (Kinchin and Mougouei 2022).
3. The impact of artificial intelligence on working employees: Overuse can lead to replacing some employees, raising concerns about unemployment.
4. Promoting prejudice: Artificial intelligence can have an impact on prejudice. If artificial intelligence systems are trained, it takes the system. Based on biased data, this bias may be transmitted to decisions that
5. Loss of control: There may be fears of losing control over the artificial intelligence system. When the system can learn his decisions will evolve in the future.
6. Accuracy and objectivity: AI systems can be vulnerable if they are trained on unbalanced or biased data, especially for siding. This can lead to making wrong or unfair decisions.
7. Privacy issues: AI systems can be vulnerable in hacking or misuse, which may result in security breaches or privacy.
8. Social Justice: Artificial intelligence can negatively affect society, exacerbating economic inequalities or discriminating against minorities.
9. Legal liability: It is still unclear who bears responsibility legal liability for damages caused by artificial intelligence systems.

There is a broad consensus on establishing rules and clearly setting moral boundaries. To this end, the Swiss government has established an interdepartmental working group on artificial intelligence. At the end of November 2020, the government adopted the guidelines for artificial intelligence developed by the team under the leadership of the federal administration for economics, education, and scientific research.

Due to the significant differences between the basic ethical values that apply in different countries, it is completely uncertain whether there will be a unified global regulation of artificial intelligence or not. Perhaps the main argument in favor of using artificial intelligence is its ability to simplify our lives. This is mainly done through its systems that help us save time and take over the completion of tedious tasks, such as the "chat GBT" ChatGPT that performs them. In the best-case scenario, artificial intelligence can save lives; for example, a rescue drone can enter very narrow areas that are difficult for ordinary rescue aircraft to reach.

6.6 Conclusion

The artificial intelligence and transformation of the international economy are nothing but indeed new technological activities, scientific knowledge associated with business practices. The more recent tremendous rise in the use of AI and especially ML has been enabled by 2 things, compute accelerators such as GPUs (graphical processing units, now also with current emerging IPUs or interactive processing units), but primarily because we have increasingly large sources of Big data which benefit from these algorithms. And this is going to have a huge impact on governments, communities, businesses and individuals as well. Finally, in these use cases stunningly high artificial agents' capabilities such as games and task recognition and classification will free space for creativity process and product innovativeness. This has resulted in creating products and technologies for people with disabilities, as well as the large elderly population. It will also allow for the development of an interactive and interesting play experience that could improve cognitive intelligence, emotional ability of children while making a contribution to toys and games industry. Overall, autonomous technologies involve Autonomous in nearly all fields and the establishment of numerous AI-based products and services that will revolutionize health, learning possibilities, security solutions for personal safety, and general well-being as well as trade among other parts your life. But there are some specific security, safety, and ethics-related concerns of using artificial intelligence technology which requires a bit focus.

Built upon existing firms, much innovation and worldwide competitive yarn is also acting on them as they will become AI corsair companies. The real ambition is to evolve with the most up-to-date AI and find out who gets there first in terms of technology. In this piece, we showed the sectors of automation and artificial intelligence with most growth potential in the near future which will impact healthcare, cybersecurity, business operations analytics as well as marketing/sales among others plus a generational increase in investment into automating and leveraging AI since 2010. This is obviously because automation and AI has the potential to change the economy. Thus, we can see that artificial intelligence is still limited to a few regions, creating an artificial intelligence gap. Moreover, further technological expansion and extension in countries enabled by artificial intelligence and the lack of participation of other countries will widen this gap, representing the current digital divide. This will deepen inequality in the economic, cultural, and social sectors and create the current gap. That is, this gap represents the dark side of artificial intelligence. Accordingly, the three-dimensional analysis has presented the state of the art of research, creativity, and the spread of the robot and artificial intelligence market. He also provided a better understanding of how artificial intelligence can reshape international markets, transform innovation processes, research and development organizations, and the functioning of business processes and the international economy. Knowledge about each of these areas will familiarize the sectors with the development and application of artificial intelligence shortly. This also helps policymakers identify challenges and discover legal and ethical concerns in artificial intelligence. Categorically, artificial intelligence will positively impact education, living standards, health, and business. However, its initiatives should be strengthened toward exploring ethical issues and reducing the AI gap between the various sectors in the country and the structures of other countries developed in artificial intelligence.

6.7 Implications and Future Directions

Artificial intelligence (AI) is an effective tool used in many fields, including the field of charity and humanitarian work. It can have a significant impact on improving the quality of life for individuals and communities in need by improving the efficiency of charitable and humanitarian organizations, and expanding their reach to those in need. Exploiting the opportunities and mitigating the risks of new and emerging technologies in humanitarian work requires adherence to a rights-based approach in accordance with international law, the principles of "harmlessness", and humanitarian principles. This approach ensures the availability of the required basics, such as data protection, privacy, and liability standards. Investment in communication and digital literacy will help to bridge digital gaps and provide assistance to local methods, while cooperation with other actors will strengthen cooperation between areas of expertise and the provision of resources.

By following "harmlessness", and humanitarian principles procedures and principles, it is possible to reduce the danger of artificial intelligence and the achievement of effective utilization and responsible for artificial intelligence technology (Pino Barranco, 2020).

Robust regulatory framework: Governments should establish strict laws and regulations to control ethical, security and privacy standards. The use of artificial intelligence, ensuring its compliance with legislation and regulation: Governments should regulate the use of artificial intelligence. By establishing strict laws and regulations to control education and awareness, education should provide appropriate training to understand and deal with application challenges and risks.

References

Al-Enezi T (2022) Artificial intelligence as an input to achieving sustainable development in the business environment in accordance with the pillars of the kingdom's vision 2030. J Econ, Adm Leg Sci 6(13):48–63

Alesani D, Nasi G, Manni S (2024) Digital transformation in IOs–toward a human-centered approach. In: Management of international institutions and NGOs. Routledge, pp 406–431

Belfar S, Fenides A, Bouktab K (2021) Using artificial intelligence to achieve development sustainable. Book of the international Maghreb conference on sustainable development updates, 323–336

Bhima B, Zahra ARA, Nurtino T, Firli MZ (2023) Enhancing organizational efficiency through the integration of artificial intelligence in management information systems. APTISI Trans Manag 7(3):282–289

Bouzid (2022) Employing artificial intelligence applications in electronic marketing. Khazzartech Ind Econ 12(01): 469–486

Cath C, Wachter S, Mittelstadt B, Taddeo M, Floridi L (2018) Artificial intelligence and the 'good society': the US, EU, and UK approach. Sci Eng Ethics 24:505–528

Chin T, Cheng TCE, Wang C, Huang L (2024) Combining artificial and human intelligence to manage cross-cultural knowledge in humanitarian logistics: a Yin-Yang dialectic systems view of knowledge creation. J Knowl Manag 28(7):1963–1977. https://doi.org/10.1108/JKM-06-2023-0458

Chowdhary KR (2020) Fundamentals of artificial intelligence. New Delhi, Springer India, pp 603–649

Dave A (2017) Digital humanitarians: how big data is changing the face of humanitarian response. Bioethical Inq 14:567–569. https://doi.org/10.1007/s11673-017-9807-8

Davydova I, Andronov I, Zhurylo S, Zubar V, Iliopol I (2023) Prospects for the use of artificial intelligence in jurisprudence: from the educational process to legal practice. The experience of China. Rev Eduweb 17(4):87–95. revistaeduweb.org

Domini B, Dewi AS, Cesna GP (2023) The effects of Artificial Intelligence (AI) on company performance including the value of ai-based business transformation initiatives. IAIC Trans. Sustain. Digit. Innov 5(1):24–38

Dubey R, Bryde DJ, Dwivedi YK, Graham G, Foropon C (2022) Impact of artificial intelligence-driven big data analytics culture on agility and resilience in humanitarian supply chain: a practice-based view. Int J Prod Econ 250:108618. https://doi.org/10.1016/j.ijpe.2022.108618

Dwivedi YK, Hughes L, Ismagilova E, Aarts G, Coombs C, Crick T, Duan Y, Dwivedi R, Edwards JS, Eirug A, Galanos V, Ilavarasan PV, Janssen M, Jones P, Kar AK, Kizgin H, Kronemann B, Lal B, Lucini B, Medaglia R, Meunier-FitzHugh KL, Meunier-FitzHugh LC, Misra SK, Mogaji E, Sharma SK, Singh J, Raghavan V, Raman R, Rana NP, Samothrakis S, Spencer J, Tamilmani K, Tubadji A, Walton P, ... Williams MD (2021) Artificial Intelligence (AI): multidisciplinary perspectives on emerging challenges, opportunities, and agenda for research, practice and policy. Int J Inf Manag 57 101994

Fernández-Luque L, Imran M (2018) Humanitarian health computing using artificial intelligence and social media: a narrative literature review. Int J Med Informatics 114:136–142. https://doi.org/10.1016/j.ijmedinf.2018.01.015

Fernandez-Luque L, Imran M (2018) Humanitarian health computing using artificial intelligence and social media: a narrative literature review. Int J Med Informatics 114:136–142

Feroz AK, Zo H, Chiravuri A (2021) Digital transformation and environmental sustainability: a review and research agenda. Sustainability 13(3):1530

Grosan C, Abraham A (2011) Advanced artificial neural networks. In: Intelligent systems reference library, vol 17. Springer, Berlin, Heidelberg. https://doi.org/10.1007/978-3-642-21004-4_13

Guerrero-Roldán AE, Rodríguez-González ME, Bañeres D, Elasri-Ejjaberi A, Cortadas P (2021) Experiences in the use of an adaptive intelligent system to enhance online learners' performance: a case study in economics and business courses. Int J Educ Technol High Educ 18:1–27

Hamid AA, Alemu AE (2023) Technology and Halal retailing sector: digital transactions and what is new? In: Technologies and trends in the Halal industry, Routledge, pp 221–232. https://doi.org/10.4324/9781003368519-19

Kadyrova GM, Shedko YN, Orlanyuk-Malitskaya LA, Bril DV (2021 March) Prospects to create and apply artificial intelligence in the activities of public authorities. In: IOP conference series: earth and environmental science, Vol, 650, No. 1. IOP Publishing. iop.org, p 012014

Khan M, Parvaiz GS, Abolghasemi A, Jehangir M, Hassan NA, Bae JH (2022) A model for understanding the mediating association of transparency between emerging technologies and humanitarian logistics sustainability. Sustainability 14(11):6917. https://doi.org/10.3390/su14116917

Khanmohammadi S, Golafshani E, Bai Y, Li H, Bazli M, Arashpour M (2023) Multi-modal mining of crowdsourced data: efficient provision of humanitarian aid to remote regions affected by natural disasters. Int J Disaster Risk Reduct 96:103972

Kinchin N, Mougouei D (2022) What can artificial intelligence do for refugee status determination? a proposal for removing subjective fear. Int J Refug Law 34(3–4):373–397

Madianou M (2021) Nonhuman humanitarianism: when'AI for good'can be harmful. Inf Commun Soc 24(6):850–868

Marelli M (2022) The solarwinds hack: lessons for international humanitarian organizations. Int Rev Red Cross 104(919):1267–1284

Martin A, Sharma G, Peter de Souza S, Taylor L, van Eerd B, McDonald SM, Marelli M, Cheesman M, Scheel S, ... Dijstelbloem H (2023) Digitisation and sovereignty in humanitarian space: technologies, territories and tensions. Geopolitics 28(3):1362–1397

Martin A, Sharma G, Peter de Souza S, Taylor L, van Eerd B, McDonald SM, Marelli M, Cheesman M, Scheel S, … Dijstelbloem H (2022) Digitisation and sovereignty in humanitarian space: technologies, territories and tensions. Geopolitics 28(3):1362–1397. https://doi.org/10.1080/14650045.2022.2047468

Maslej N, Fattorini L, Brynjolfsson E, Etchemendy J, Ligett K, Lyons T, Manyika J, Ngo H, Niebles JC, Parli V, Shoham Y, Wald R, Clark J, Perrault R, ... Perrault R (2023) Artificial intelligence index report 2023. arXiv preprint arXiv:2310.03715

Zembylas M (2023) A decolonial approach to AI in higher education teaching and learning: strategies for undoing the ethics of digital neocolonialism. Learn Media Technol 48(1):25–37

Muhammad H (2020) Artificial intelligence and its impact on economic activity. Al-Hikma J Educ Stud Psychol 7 4:8

Pino Barranco B (2020) International responsibility of states and artificial intelligence. Unpublished masterthesis, CEI, Centro Adscrito a la Universitat de Barcelona.https://www.diposit.ub.edu/dspace/bitstream/2445/170430/1/TFM_Beatriz_Pino.pdf

Rahman NSFA, Hamid AA, Lirn TC, Al Kalbani K, Sahin B (2022) The adoption of industry 4.0 practices by the logistics industry: a systematic review of the gulf region. Clean Logist Supply Chain 5:100085. https://doi.org/10.1016/j.clscn.2022.100085

Sachdev K, Mehra A (2022) An exploration of philosophical ideas of select thinkers in artificial intelligence. Int J Soc Sci Hum Res 05(12). https://doi.org/10.47191/ijsshr/v5-i12-31

Tomaszewski B (2008) Producing geo-historical context from implicit sources: a geovisual analytics approach. Cartogr J 45(3):165–181

Ülkü MA, Bookbinder JH, Yun NY (2024) Leveraging industry 4.0 technologies for sustainable humanitarian supply chains: evidence from the extant literature. Sustainability 16(3):1321

Van den Homberg MJ, Gevaert CM, Georgiadou Y (2020) The changing face of accountability in humanitarianism: Using artificial intelligence for anticipatory action. Polit Governance 8(4):456–467

Wang L (2018) Artificial intelligence: transforming industries. Harvard business review. https://www.hbr.org/artificial-intelligence-transforming-industries

Dr. Abdelsalam Adam Hamid he is an assistant professor in Supply chain Management at A'Sharqiyah University, Sultanate of Oman. He served as an assistant professor of supply chain at International Maritime College Oman. Prior to that he worked as an assistant professor and head of Business Administration Department at Sudan University of Science and Technology also he was a researcher at Hanken School of Economics, Finland. Abdelsalam has published books, chapters, and more than 20 articles in peer-reviewed journals. His research interests lie in supply chain strategy, practices, logistics operations management, and industrial marketing.

Abebe Ejigu Alemu is logistics management associate professor in the Department of Logistics Management, International Maritime College Oman, National University of Science and Technology; professor of Marketing and Supply Chain Management, School of Management, Mekelle University. He did his Post-Doc Research Fellow: SARChI chair for social policy, PhD in Marketing and supply chains management (KU Leuven, Belgium). He taught several courses in the department of management at BA, MBA, MA in logistics and supply chain management, MA in marketing, and PhD in Management studies. He also served the university by assuming various university positions such as Vice President for Support Services (Admin. Bus. and Development); Director, Academic Programs Directorate; Head of Department of Management; MBA program coordinator; Head, Higher education reform office, Mekelle University; Head, Human Resource Development Department; Head, Department of Management. He has contributed more than 35 research publications presented in international and national journals and presented several papers in international and national conferences and workshops.

7 Empowering Humanitarian Efforts: IoT-Enabled Solutions for Distribution Center

Muhamad Munzir Khairuddin
and Nor Aida Abdul Rahman

Abstract

In the face of natural disasters, conflicts, and other humanitarian crises, the efficiency of distribution center in distributing the aids supplies is crucial for affected communities. This disaster will impact humans' daily life, the victims will suffer and have difficulties to fulfill their daily need such as food and beverage, medication, hygiene need, sleep requirement, etc. Humanitarian aid effort is one of the initiatives to channel the human essentials to the victim in order to sustain their life especially for woman, kids, and patients. Humanitarian aid effort looks difficult and challenging with current conventional operational without any use of technology. The application of technology is deemed important in humanitarian aid effort operation to improve the aids supply and deliveries to the victims. The type of technology can be applied focus in the distribution center such as IOT devices like censors, RFID, Camera, etc. The implementation and integration of these technology kits will improve the humanitarian aids effort operation as to make sure the disaster victim not really impacted to their mental and physical health.

Keywords

Humanitarian effort · Technology enable · IOT distribution center · Crisis management · Technology adoption

7.1 Introduction

Natural disaster is unwanted tragedy to everybody, but it is uncontrolled by anyone. The natural disaster like flooding, drought, earthquake, and so on including man-made disaster from war crisis will impact the human daily life freedom. The essential life needs for human such as food and beverage, medication, toiletries, sleep requirement like bed sheet, blanket, and so on will be unavailable. In order to deliver those human needs to victims, humanitarian aid effort is most indispensable to make sure the kids, woman, and patients can still survive during this harsh critical time (Altay & Labonte 2014).

Currently most of the humanitarian aids manage and operate conventionally via distribution center, but negatively they are ineffective and inefficient for long run and involve with big

M. M. Khairuddin (✉) · N. A. A. Rahman
Universiti Kuala Lumpur, Malaysian Institute of Aviation Technology,
Subang, Selangor, Malaysia
e-mail: mmunzirk@unikl.edu.my

N. A. A. Rahman
e-mail: noraida@unikl.edu.my

N. A. A. Rahman et al. (eds.), *Supply Chain in Humanitarian Operations*, Sustainable Development Goals Series,
https://doi.org/10.1007/978-981-96-4426-1_7

scale of victims. In humanitarian aid effort, most crucial service need to be available is aid management and delivery by transportation modes; the mode of transportation can be by air transportation, water transportation, or ground transportation. The distribution center is an essential place in humanitarian effort operation. Why a distribution center is important in humanitarian aid effort operation? The answer is because, it is the beginning or first phase of this operation where the collection of essential goods and products are placed and collected before they are distributed to the victim. There will be a few processes that need to undergo, before the delivery and distribution can be made. Currently, the method applied is not efficient enough to cater and manage the supplies since the demand from impacted disaster victims is highly demanded from many aspects like foods, drinking water, medication, hygiene needs, sleeping needs, and many more. The supplies must be well managed to make sure the essentials are ready stock but need for proper communication and coordination from suppliers and vendors. The storage and distribution center will be messy and super busy without proper planning and managing. In order to improve the effectiveness and efficiency of the storage and distribution center, the adoption and implementation of technology is deeply important to improve the humanitarian aid operation effort in distribution center. The rise of the Internet of Things presents a promising avenue for enhancing the capabilities of humanitarian aid storage and distribution centers. IoT-enabled solutions can streamline logistics, improve inventory management, and facilitate real-time monitoring of supply chain operations, ultimately enhancing the speed and precision of aid delivery (Kadar et al. 2018).

The technology kits like RFID, censors, camera, and so on are very useful, but if they are operating individually without integration with each other, there will be less productivity. The Internet of Things (IoT) systems can provide real-time visibility into stock levels, expiration dates, and the condition of stored goods (Khanna and Tomar 2016). This allows aid organizations to proactively address potential shortages, minimize waste, and ensure that supplies reach their destinations before they become unusable. Furthermore, IoT-enabled distribution centers can leverage connected devices to optimize routing, monitor transportation, and coordinate the movement of goods more efficiently (Kadar et al. 2018). Hence, this paper will propose the technology integration model for distribution center to improve the operation of humanitarian aid effort.

Problem Statement

Humanitarian aids operation is crucial and highly important to oversee. It needs for proper plan for urgent delivery in order to save the lives of disaster victims. This operation needs for coordination from many parties and agencies to make sure the operation is smooth and without any delay. The type of party may involve in this aid humanitarian operation for example goods and products suppliers, transportation operators, warehouse operators, and so on. Same goes to some of government agencies which need during the humanitarian aid operation such as National Security Council, military department, fire fighters, medical service, etc.

During this urgent and crucial time, coordination of plan must be super-efficient, in order to reduce death, injuries, and starvation. The location of facility like command center, storage or distribution center must be decided as soon as possible. There were few of disaster reported have taken longer time for their rescue service delivery because of lack of technology and logistical infrastructure. The choice of facility location must be nearest to the disaster location, safe and assessable by any modes of transportation. The distribution center must be equipped with high technology equipment and kits, so that the information can be process as soon as possible and the deliveries of aids can be made as soon as possible. Humanitarian efforts in disaster-stricken areas are often hampered by delays in the delivery of essential supplies due to a lack of advanced technological integration within storage and distribution centers. This issue has been highlighted in various reports and real-life scenarios. Below were 4 cases which involve of delay on their humanitarian effort delivery because of technology adoption lacking;

(1) In the aftermath of the 2017 Kermanshah earthquake in Iran, humanitarian agencies struggled with timely delivery of aid due to inadequate technology and logistical infrastructure.
(2) During the humanitarian response to the 2010 Haiti earthquake, the absence of advanced technological systems resulted in severe delays and mismanagement of resources, impacting the efficiency and speed of aid distribution.
(3) During the COVID-19 pandemic, many logistical operations faced significant delays because of outdated technologies and inefficient data management systems, resulting in critical shortages and slow responses to the needs of affected communities.
(4) Department of Defense (DOD) has encountered challenges with its digital logistics systems, where outdated technology and lack of system integration led to vulnerabilities and inefficiencies in supply chain operations, which could be detrimental during disaster relief efforts.

Above are the real scenarios impacted because of lack of technology application during humanitarian aid effort operation. The effect from technology lacking may involve loss of soul and life of disaster victims. Hence, by these situations showing us on the importance of the technology application during humanitarian effort and disaster relief effort specifically in distribution center. In addition, there are very critical of research number on the technology application for humanitarian effort operation specifically in the area of distribution center. Table 7.1 highlights the list of research papers focusing on the type of technology and IOT application on humanitarian effort for warehouse and distribution center.

Most of the research papers focus on the IoT application and has not captured the whole integration of technology to be fit in the distribution center which relates to this statement. This is the gap for this area where the function of the technology has not really utilized and not really provided full information toward operator. As mentioned in Abdul Rahman et al. (2022), the lack of technological integration in many humanitarian operations often hampers the delivery of vital resources to those in need. Hence, this paper will propose the technology integration model for distribution center in order to improve the operation of humanitarian aid effort operation.

7.2 Research Objectives

The research objectives of this study are;

(i) To explores the type of Internet of Things (IoT) technologies in enhancing the efficiency and effectiveness of humanitarian operations.
(ii) To suggest an IOT integration model for humanitarian effort storage and distribution center

This study is to investigate the types of IoT and will suggest the technology integration model for distribution center which can revolutionize humanitarian effort distribution centers by providing real-time data, improving inventory management, and ensuring aids can be deliver effortless.

7.3 Literature Review

Activity Inside Humanitarian Effort Distribution Center

The most essential facility during humanitarian aid operation is distribution center. This facility will support the most important process during humanitarian aid operation which are receiving until delivering process. Receiving can be highlight on the process of receiving the supply aid from supplier or vendors or donators,

Table 7.1 List of research papers focusing on IoT application on logistics and warehouse

Research Paper	Year	Authors	Abstract
Optimizing Humanitarian Logistics Using IoT and Data Analytics	2021	M. Brown, L. Davis	This study explores the use of IoT and data analytics to improve logistics in humanitarian operations, with case studies on disaster relief and food distribution
The Role of Blockchain and IoT in Humanitarian Supply Chains	2022	D. Garcia, N. Patel	This paper examines the synergy between blockchain and IoT technologies in enhancing transparency and efficiency in humanitarian supply chains
Application of Internet of Things (IoT) in Sustainable Supply Chain Management	2023	Syed Fayaz Ahmad, Ahmad Y. A. Bani Ahmad, Nasir Khan	This article examines how IoT technologies can enhance supply chain efficiency and reduce costs. It discusses smart warehousing and secure distribution of products, emphasizing sustainable supply chain management
Analyzing Barriers to Internet of Things (IoT) Adoption in Humanitarian Logistics: An ISM–DEMATEL Approach	2024	Abderahman Rejeb, Karim Rejeb, Imen Zrelli	This study explores barriers to IoT adoption in humanitarian logistics, identifying and classifying 12 key obstacles. It highlights the importance of overcoming these barriers to enhance efficiency and decision-making in relief operations
Internet of Things (IoT) in Humanitarian Supply Chains: A Framework for Improved Coordination	2024	Syed Fayaz Ahmad, Ahmad Y. A. Bani Ahmad, Nasir Khan	This paper presents a framework for how IoT can improve coordination in humanitarian supply chains, with a specific focus on warehouse and distribution center operations
Leveraging IoT for Humanitarian Aid: Enhancing Warehouse and Distribution Efficiency	2024	Syed Fayaz Ahmad, Ahmad Y. A. Bani Ahmad, Nasir Khan	This study highlights the benefits of IoT in humanitarian aid, specifically improving the efficiency and accuracy of warehouse and distribution center operations

while delivering can be highlight on the process of delivery the supplies or aid or resource to the disaster victims after other few activities like segregating, grouping, and so on. These are the crucial processes which need a lot of space and volunteers to make sure the aids are evenly packed before the delivery is made. See Fig. 7.1 and Table 7.2 for details. Figure 7.1 shows the process flow of the humanitarian effort activities at the distribution center, while Table 7.2 explains the description for each activity.

The activity in Humanitarian effort distribution center:

Type of Technology

Technology is the application of scientific knowledge for practical purposes, especially in industry. It is an innovation of technology to improve the outcome of the work, for example, the innovation of car, focus to improve the movement of people from origin to destination. Internet of thing (IoT) enables numerous devices to remain connected to the network permanently transmitting the data without interruption (H. Xu et al. 2013). The kits may include under IoT device censors like surveillance camera, RFID reader, thermometer, movement censor, and so on.

In the humanitarian effort in distributing the aids to victims of disaster, the use of technology must start at the storage or distribution center at first, it is because when the goods or products receive from vendor or supplier or donators, the distribution process will be faster and efficient.

The function of technology proposedly to help manage the data related with goods and products received at distribution center, for example by using RFID, the data from goods received can be directly record into the data base for example the type of product, quantity, expiry date, manufacturer information, and much more. These information in very important

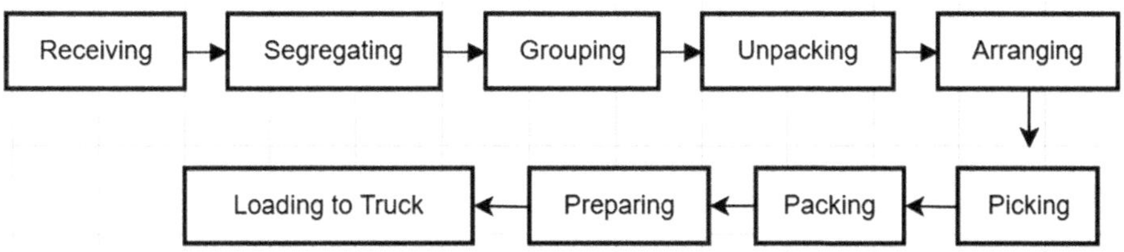

Fig. 7.1 Humanitarian effort distribution center process flow

Table 7.2 Type of activities and task description inside humanitarian effort distribution center

Type of activity	Task Description
Receiving	Accepting and verifying the delivery of goods from suppliers, vendor and donators
Segregating	Sorting received goods based on categories, types, or other criteria
Grouping	Collecting similar items together for further processing
Unpacking	Removing items from their packaging materials
Arranging	Organizing items in a systematic manner for easy access
Picking	Selecting items from storage area to be pack
Packing	Wrapping and packaging selected items for delivery
Preparing	Getting items ready for loading, which may include labeling
Loading to truck	Placing the prepared packages onto trucks for delivery

to manage and classify those products into category. The data were transmitted directly into database automatically via RFID reader without any delay and missing. Utilizing wireless sensor networks and RFID technologies, IoT systems can provide real-time visibility into stock levels, expiration dates, and the condition of stored goods (Khanna and Tomar 2016). There were other types of devices can be adapted into humanitarian effort distribution center. The list of the technology device with description as per table 7.3:

Via integration and adoption of technology mention above, the goods and supply can be manage efficiently and effectively. The goods and supply can be traced and detected only using the application software at real time on the status of supplies update. Figure 7.2 illustrates the process flow for the technology application for the humanitarian effort distribution center.

The diagram below shows the linkage of technology integration among them to improve the distribution center roles on its duty to manage the supplies and deliver the aid to the disaster victims. Every process is recorded using specific technology into the data base. When suppliers or vendors or donators arrive for their supply, the supply will be recorded via RFID or barcode, the data will record into database system using Inventory Management system. The integration of IoT and blockchain technology for inventory management system will allow the process of information collection provided by devices in decentralized manner, at the same time reducing the cost related to the large data center maintaining cost (H. F Atlam et al. 2018).

The IMS not only focuses on the aid supply data but extends the data into block chain system including the recording of surveillance system. Block chain is a recorded transaction in correct sequence, it made sure no alteration can be made to the information and manipulated (M. Dobrovnik et al. 2018). The blockchain is a new paradigm in term of trust in distributed

Table 7.3 List of technology types, function, and task

No	Technology type	Function	Tasks
1	Inventory Management System (IMS)	Software solution for managing and tracking inventory in real-time	– Real-time inventory monitoring – Automated replenishment – Reporting and analytics – Demand forecasting
2	Blockchain	Provides a decentralized, immutable ledger for recording transactions	– Supply chain transparency – Authentication and verification of supplies – Real-time updates – Disaster response coordination
3	RFID (Radio Frequency Identification) and RFID reader	Uses electromagnetic fields to identify and track tags on objects	– Real-time tracking of supplies – Automated scanning of multiple items – Access control for sensitive areas - RFID reader read the data transmit from RFID tag
4	Barcoding	Uses unique barcodes printed on items that can be scanned	– Efficient inventory management – Tracking and tracing of items – Batch management for the aid package
5	Surveillance Cameras	Monitors physical security and operations within the center	– Security monitoring to prevent theft and unauthorized access – Operational oversight to ensure efficiency – Compliance and safety monitoring – Incident documentation

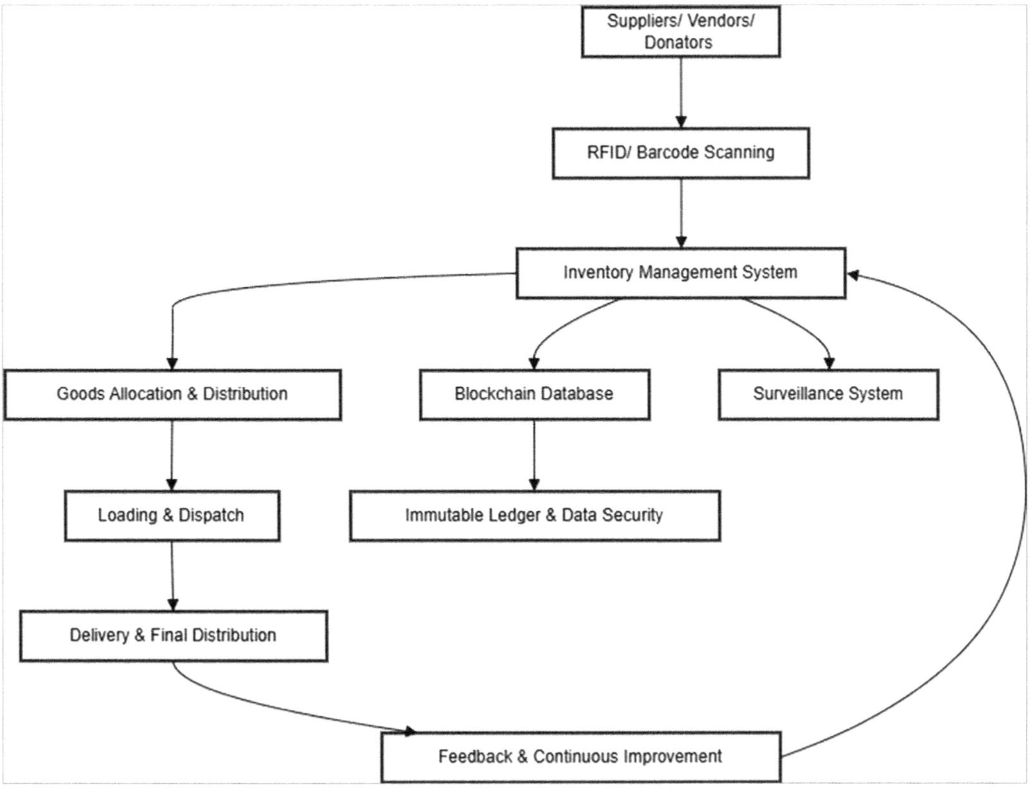

Fig. 7.2 Technology integration model for distribution center

information in any transaction (Aranda et al. 2019). The goods and supply received then process and packing based on the allocation and where to distribute target based on the disaster victim recorded.

The number of supply pack will be loaded and out to dispatch after recording their number of aid pack to the disaster location. Lastly, delivery and final distribution is made to the disaster victims efficiently without any waste. The data of delivery made can be uploaded to the IMS system for the feedback and improvement.

7.4 Conclusion

In the humanitarian effort operation for disaster is not only to recover and secure the victims of disaster, but also need to focus on the plan to mitigate the aid supplies which need to be distributed to the victim. The aid supply firstly receives by distribution center as a place of supply management.

The distribution center plays a crucial role as the point where supplies are received, sorted, and loaded onto trucks before being distributed. In conventional systems, this process can slow down aid delivery to victims; however, adopting technology in distribution centers can enhance supply management and significantly improve distribution and delivery efficiency.

The technology will benefit less if it stood alone and operated individually, by integrating specific devices into the inventory management system (IMS); it could be a systematic system to improve the operation of humanitarian effort. The whole operation will be smooth, efficient, and effective to manage since the data can be reached and reviewed via database. The aid supply can be managed until delivery and distribution is made without any waste to the specific disaster victims. Lastly the integration of IoT in humanitarian aid storage and distribution centers is not without its challenges. Ensuring the reliable and secure transmission of data, addressing privacy concerns, and overcoming the lack of digital infrastructure in many crisis-affected regions are just a few of the hurdles that must be navigated (Khan et al. 2020; Shah et al. 2019; Al-Rawashdeh et al. 2022).

References

Abdul Rahman NA, Ahmi A, Jraisat L, Upadhyay A (2022) Examining the trend of humanitarian supply chain studies: pre, during and post COVID-19 pandemic. J Humltarian Logist Supply Chain Manag 12(4):594–617. https://doi.org/10.1108/JHLSCM-01-2022-0012

Al-Rawashdeh M, Keikhosrokiani P, Belaton B, Alawida M, Zwiri AMA (2022 July 19) IoT adoption and application for smart healthcare: a systematic review. Multidiscip Digit Publ Inst 22(14):5377–5377. https://doi.org/10.3390/s22145377

Aranda DA, Miguel Molina Fernández L, Stantchev V (2019) Integration of Internet of Things (IoT) and blockchain to increase humanitarian aid supply chains performance. In: The 5th international conference on transportation information and safety, July14–July 17, 2019, Liverpool, UK

Xu H, Liu W, Wang C, Chen H (2013) Service oriented management for internet of things. International Journal of Information and Network Security 2(1)

Atlam HF, Alenezi A, Alassafi MO, Wills GB (2018) Blockchain with internet of things: benefits, challenges, and future directions. Int J Intell Syst Appl 10(6), 40–48

Kadar HH, Rafee PAA, Sameon SS (2018). Internet of Things (IoT) and water crisis. In: 2018 4th International Conference on Computer and Information Sciences (ICCOINS), pp 1–6

Khan S, Wani MM, Alam M (2020 January 1) IoT applications in urban sustainability. Cornell University. https://doi.org/10.48550/arxiv.2008.10656

Khanna A, Tomar R (2016) IoT based interactive shopping ecosystem. In: 2016 2nd international conference on Next Generation Computing Technologies (NGCT), 29, 40–45.

Dobrovnik M, Herold DM, Fürst E, Kummer S (2018) Blockchain for and in logistics: what to adopt and where to start. Logistics 2(3):18

Altay N, Labonte M (2014) Challenges in humanitarian information management and exchange: evidence from Haiti. Disasters 38(s1):S50–S72

Shah STU, Yar H, Khan I, Ikram M, Khan H (2019) Internet of things-based healthcare: recent advances and challenges. In: EAI/springer innovations in communication and computing. Springer International Publishing, pp 153–162

Muhamad Munzir Khairuddin is currently a researcher and full-time lecturer at Universiti Kuala Lumpur, Malaysian Institutes of Aviation. He has a background of aviation industry education from his Bachelor degree. He earned his Master degree in Management from Universiti Kuala Lumpur, Business School. His master research titled "An Investigation towards the Factors Impeding the Success of Halal Compliance in Air Cargo Warehouse in Malaysia". Currently, he is in progress to complete his PhD in Management.

Nor Aida Abdul Rahman currently serves as an associate professor at Universiti Kuala Lumpur, Malaysian Institute of Aviation Technology (UniKL MIAT) in Subang, Selangor, Malaysia. Her work has been featured in prestigious journals such as Industrial Marketing Management, Journal of Islamic Marketing, Journal of Humanitarian Logistics and Supply Chain Management, and Journal of Quality and Reliability Management, among others. In addition to her role at UniKL MIAT, she holds visiting appointments as a professor at ITL Trisakti in Jakarta, Indonesia, and as a visiting fellow at the UKM-MPOB Endowment Chair at the National University of Malaysia. Dr. Nor Aida has a distinguished record of scholarship, having authored or co-authored over 100 articles, book chapters, and research books. Her expertise also extends to the publishing realm, where she has contributed to the editorial teams of more than 10 books with Routledge and Springer. She earned her PhD in Management (Supply Chain Management) from Brunel University, London, after completing her Master's and Bachelor's degrees in Business and Administration from the National University of Malaysia. Her dedication to the field is further demonstrated by her roles as an MQA panel auditor, a JAKIM halal-certified trainer, an HRDF-certified trainer, and a Board Member of Education at the Chartered Institute of Logistics and Transport Malaysia (CILTM).

How Blockchain Technology Manages Supply Chain Risk

8

Jeremy Ephraim Simarmata

Abstract

Supply chain risk is present and looming over all supply chains. Supply chains, by nature, are entities with many distinct stakeholders all of whom have their own innate risks. Lowering, containing, and preparing for all forms of supply chain risks are mission-critical to many organizations, business, or otherwise. This chapter will dive into the world of supply chain risks and how a promising technology, blockchain, may be the answer to the risks currently faced by the world's supply chains. This chapter will first thoroughly examine the types of risks present within supply chains, their impacts, and current strategies and technologies used to mitigate them. Then it will examine the feasibility of blockchain technology in addressing these risks. It is evident that supply chains can be the source of competitive edge for many organizations; it is important that new technologies be examined and implemented into current organizations.

Keywords

Supply chain · Risk · Blockchain · Risk mitigation · Humanitarian supply chain

8.1 Introduction

Supply chains are the heart of every business, industry, and organization. It pumps the raw materials, goods, and everything else to where it needs to be. The recent pandemic has shown that a well-designed and managed supply chain is a reliable tell of a competitive and thriving company (Francis et al. 2021). Many of the most important supply chains in the world span across the globe, receiving raw material parts from one part of the world, manufacturing it in another, then selling said product in yet another part of the world. These activities enable and propel economic activity and encourage collaborative behavior (Baah et al. 2022).

As an intricate chain of suppliers, manufacturers, and many other parties as well as layers of said parties involved, it is expected that supply chains have risks. These risks range from a slight setback that can be undone momentarily to an existential threat to an organization. Serious risks can stem from many parts of the supply chain and it is important for organizations to put an emphasis on mitigating and

J. E. Simarmata (✉)
Department of Business Management, Trisakti Institute of Transportation and Logistics, East Jakarta, Indonesia
e-mail: jeremyephraim05@gmail.com

N. A. A. Rahman et al. (eds.), *Supply Chain in Humanitarian Operations*, Sustainable Development Goals Series,
https://doi.org/10.1007/978-981-96-4426-1_8

controlling these risks (Domnina et al. 2022). Largely, these risks are classified into four main groups: operational risks, financial risks, compliance risks, and strategic risks.

Firstly, operational risks are related to the day-to-day internal operations of the supply chain. These risks may involve things such as logistics disruptions, production delays, and machine breakdowns. Operational risks can have a serious impact on supply chains. It has implications such as the breach of contract for late deliveries which may lead to the lack of fulfillment; this one example already includes customer dissatisfaction, increased operational costs, and unrealized profits (Nimmy et al. 2022).

Next, financial risks are centralized around the potential monetary losses that occur due to price volatility, fraud, theft, and supply chain partners' breach of contract. Businesses tend to focus on building defenses against financial risk in their supply chain as they can be detrimental in a short amount of time. A common issue that many companies face is the supply chain partners failing to deliver what was agreed upon. If a key supplier were to fail to supply, it could cause a significant amount of financial damage for said company (Sinaga et al. 2019).

Third is compliance risk. Compliance risks are concerning the compliance with laws and regulations that are in place for the supply chain. Most supply chains are global, so it is important to adhere to regulations, policies, and laws that govern. Failure to comply can result in a number of different ways, namely fines, penalties, and damaged reputation. Fines and penalties can be administered by the respective governing bodies, whereas the damaged reputation can have further implications. A damaged reputation because of unethical sourcing for a brand whose target audience is people with an interest in sustainability and equitable business practices could destroy the brand completely (Kusrini and Hanim 2021).

Lastly, strategic risks. Strategic risks come from decision-makers. Decision-makers in an organization can set a strategy that has major repercussions. It may come in the form of switching to a different supplier or service provider or entering a new market. These decisions typically take a significant amount of resources and effort, and are difficult to renege without taking on substantial losses (Kramer et al. 2022).

All four of these risks are present in every supply chain. The sheer size of supply chains guarantees this. However, it does not mean that it is unavoidable or impossibly difficult to mitigate. Although mitigation will not wholly subvert the consequences, it can make these risks much more manageable during and after crises. Blockchain technology is a technology popularized by the rise of cryptocurrency. Its main offer is the feature of transparency, with many features that can be built on top of this transparency. In the case of addressing and managing supply chain risk, blockchains may be able to reduce risks such as fraud, misappropriation of funds, and many others to be reviewed in this chapter (Pattnayak and Patnaik 2022).

8.2 Research Context and Method

To collect information for this chapter, a systematic review of the literature was conducted, including academic articles, reports, and news articles related to the use of blockchain technology in addressing supply chain risks. Data from industry organizations such as the Organisation for Economic Co-operation and Development (OECD), Association for Supply Chain Management (APICS), Gartner, and the World Bank were also examined.

8.3 Literature Review

Supply chains and their risks have been thrust into the spotlight with the arrival of the COVID-19 pandemic. During this time, the impacts of supply chain disruptions materialized in a way that was felt by every consumer. Companies were feeling the pressure of higher costs or sometimes they did not have access to the materials or manufacturing they needed, which led to the higher costs being passed to the customers. It was a difficult time for the economy in general but emphasizes the importance of the supply chains' resilience (Pujawan and Bah 2022). In response to this pressure, supply chain executives had to take an objective look at their respective supply chains, finding that most of them will look increase investment in supply chain technologies with retraining their workers in order to work complementarily with the technology. Although it may seem as if technologies stopped investing in technologies, however, EY's study showed that 92% of the top companies they surveyed (over $1B revenue) did not stop technology investments, signaling their realization that the short-term problem of COVID would not trump the long-term need to compete (How COVID-19 impacted supply chains and what comes next | EY—Global, n.d.).

Blockchain technology has also risen to prominence in the past decade. It has many of the features that beneficiaries of a supply chain want. Numerous companies have started the implementation of blockchain technology into their supply chain, investing a considerable amount of their resources into this up-and-coming technology. The reasons also vary, but they usually stay within the realm of traceability, accountability, and fraud prevention (Beck 2018).

8.4 Supply Chain Resilience

Facing a more competitive environment and subject to uncertain geopolitical, as well as natural conditions; companies cannot rely on external conditions as a basis for their growth. A strong and resilient supply chain needs to be the norm and not the exception as companies with low resilience will not be able to weather disruptions as efficiently as supply chains that have been designed to survive and deliver during difficult conditions.

There are multiple ways to measure the resilience of a company's supply chain, a reliable way to measure the resilience is by using the supply chain velocity. The supply chain velocity takes into account the speed at which every supply chain process takes. Supply chain resilience was found to be a trait possessed mainly by bigger retail firms; these firms have taken well-calculated precautionary steps to ensure their supply chain is resilient. A big part of this is broad innovation. It may take the form of an extra layer of suppliers for a certain product that is vital to revenue, or taking additional measures to make sure their current suppliers are not at any risk and maintaining said relationship (Ozdemir et al. 2022).

8.5 Risk Mitigation

Broadly, risk mitigation in supply chains refers to the minimization of disruptions and the impact of disruptions. These disruptions can be detrimental to the organization if not managed and addressed properly. Risk mitigation is typically done under a set guide of best practices. Including but not limited to practices such as sourcing and supplier selection, supplier risk analysis, etc. (Kraude et al. 2022). As mentioned, the four types of risks occur due to different things, impact different things in an organization, and have different best practices currently used to mitigate and minimize their negative impact.

Operational risks are related to internal processes in an organization; thus, the mitigation strategy includes implementing strong standard of procedures, regularly maintaining assets, and having an organization with cross-functional teams. Operational risk includes issues such as logistics disruptions, supplier failures, natural disasters as well as cyber-security threats (Ríos et al. 2019).

Financial risks are more straightforward. Financial risks can also contain compliance risk as they are correlated, and finances are subject to audits. Financial risks can be mitigated by doing regular financial assessments, and eliminating inefficiencies found, hedging, and implementing audits as well (Ghadge et al. 2021).

Compliance risk can be more convoluted as they relate to legal boundaries and trade policies. Compliance risks can almost be wholly eliminated by cultivating a culture that discourages and disincentivizes lack of compliance, regular compliance training as well as a compliance system to digitalize and track.

Strategic risks can arise from top management or decision-makers. Strategic risk can be mitigated by having an agility-focused strategy that can adapt to market conditions and competitor strategies as they unfold rather than fully committing to a strategy, unwilling to change or adapt (Abrudan et al. 2022).

8.6 Background of Block Chain Technology

Blockchain technology came into existence along with the launch of Bitcoin, created by under the pseudonym Satoshi Nakamoto. Blockchain was the technology underlying Bitcoin, which facilitated its ledger system across a network of computers. Having developed over the years, blockchain technology evolved to support other applications and was implemented in various industries (*History of blockchain | Technology | ICAEW*, n.d.).

Blockchain technology is a decentralized digital ledger system. Its uniqueness comes from its immutable nature meaning that once data is recorded in the ledger, it can never be changed, altered, or deleted. The usage and value-added that this technology lends to supply chains are transparency, traceability, and security (Lewis 2015).

Blockchain technology in the supply chain is no longer strange. There have been some early adopters of this technology, typically companies that place high value on tracing their raw materials as it may be a unique selling point of said company's product. Large-scale adoption is yet to happen, as larger companies will need to see the advantages and proof of concept before adopting this technology and investing resources to make sure blockchain can work complementarily with their current infrastructure (Taherdoost 2022).

8.7 Blockchain and Supplier Relationship Management

The supply chain's reliance on suppliers is intrinsic. It is universal, ranging from e-commerce supply chains to retail, and much further. Relationships between suppliers and buyers must be mutually beneficial. Blockchain technology can boost trust in buyer and supplier relationships by increasing transparency. The mechanism by which blockchain technology will do this is its digital, immutable ledger. Not only can the movements of products and transactions can be traced and kept in this ledger, but supplier credentials and performance history can also be safeguarded on the chain (Liang et al. 2024).

As a result of the immutable record keeping, supplier and buyer disputes can be readily resolved. An issue in dispute resolution that often occurs is inconsistent records across multiple places of storage. When there is no question regarding the data and no possibility of tampering with records, it will be much easier for these relationships to mature into gainful relationships for all parties involved (Gabuthy 2023).

8.8 Cost Implications of Implementing Blockchain

Implementing a new technology is often costly, blockchain is not an exception. There are four main cost centers to review when considering upgrading to a new technology (Implementing New Technology, n.d.).

First, there are the set-up costs. This cost mainly stems from the infrastructure meaning the software and hardware needed, blockchain development costs to make sure the blockchain is according to the organization's needs and purpose, and lastly consulting services from experts to make sure the hefty investment is on the right track technically and financially. Although it may be tempting to fully utilize in-house teams and not hire consultants, the risk of setting the technology up wrong can prove to be more resource-intensive than investing in expert help in the first place.

In addition to that is operational costs. Operational costs encompass anything in the maintenance and upgrades, along with energy consumption. Maintenance of a new technology can be quite expensive as it tends to need specialized workers who have had experience with this technology beforehand. A scarcity of trained and experienced workers in new technology is the reason for the high operational costs, however, the high costs do not last for a significant amount of time as the widespread adoption of technology begets a higher number of experts in a given field, ultimately saturating the market for said experts and lowering the cost.

Third, there are training and adoption costs. Talent and human resources are crucial. Upskilling human resources and recruiting new team members is a cost center for every organization, but during times of adoption of a new technology, these costs can rise quite significantly. Apart from one-time costs for the training of employees, there is also a drop in productivity while the employees are receiving their training. Although the cost of training might not seem significant, there is a hidden cost found in the drop in productivity.

Lastly are compliance costs. Compliance costs can be attributed to the legal fees accumulated through consultation and compliance audits. These costs are necessary as failure to comply with regulations often costs more in addition to the damaged reputation of the organization. Especially in new technology such as blockchain, with the smart contract feature, it can have some intersections with legally binding documents between multiple parties. Ensuring that the legal aspect of implementation and day-to-day operations are compliant with regulations is critical in mitigating any added legal fees.

8.9 Regulatory and Compliance Aspects of Blockchain in Supply Chains

Navigating the complex legal world is a daunting challenge, add technology into the mix and it gets even more convoluted. Blockchain technology as previously discussed will have compliance costs as one of its cost centers. There are multiple concerns with blockchain technology that require it to be regulated. The main concerns are as such: Data privacy, legal enforceability, and cross-border regulations.

Data privacy is a big concern due to the prevalence of technology giants and the use of smart devices being the norm. The usage of smart devices and services comes with the price of leaving oneself vulnerable to breaches in data privacy. It has almost become a norm for these technology companies to have more relaxed data privacy policies unless public scrutiny demands their data to be protected. Blockchain's premiere feature is its transparency, this can also be a crack in its security as it is open for the whole world to see. Organizations with the intention to invest in blockchain should find a solution best tailored to their specific situation (Zhang et al. 2021).

Legal enforceability and cross-border regulations are intertwined. Legal enforceability is a risk that comes from blockchain's smart contract feature which can deliver numerous positive improvements to its users. Allowing contracts to be safer for both parties, not allowing any changes once the contract has been created, and ensuring all parties fulfill their obligations before it executes. Currently, not all jurisdictions accept smart contracts as a legally binding document, leaving room for unfair practices. Cross-border regulations fall into the same category of risk, being highly correlated with places where these contracts must be executed (Giancaspro 2017).

8.10 Interoperability of Blockchain in Global Supply Chains

Interoperability is a crucial factor in technologies in business as it precedes a complete migration. A business' legacy system would typically run on a completely different technology, form factor, and software making the interoperability of a new technology an important decision-making variable.

Blockchain facilitates efficient management of data and assets within the global supply chain by providing secure and immutable records. Automation in processes by using features such as smart contracts can streamline processes such as payments, contracts, and compliance checks.

Although its potential benefits are promising, achieving interoperability remains a difficult objective. Technical standards for data formats, protocols for seamless cross-chain communication, and standardized mechanisms need consensus before the technology's full potential benefits can be utilized. The lack of technical standards in technology is a red flag that it is not fully ready for widespread use (Zhang et al. 2020).

8.11 Environmental and Social Impact Transparency

With increasingly stronger demand for social and environmental responsibility, blockchain technology offers robust mechanisms for tracking and verifying environmental impact throughout supply chains. With its built-in immutability, blockchain ensures transparency regarding important information such as carbon emissions, waste management, and resource consumption. Transparency offers benefits to both organizations and their communities, the organizations receive value in the form of ease of tracking and monitoring their sustainability (Sansone et al. 2023).

Potential benefits include the reduction of data tampering and fraud, and enhancing the reliability of information. Transparency helps companies identify and mitigate the risks associated with non-compliance, and reputational damage (Semenzin 2023).

Although promising, integrating blockchain for environmental and social transparency remains challenging for numerous reasons, namely the need for standardization and questions of scalability.

8.12 Comparative Analysis of Blockchain and Other Risk Management Technologies

Blockchain is not the only promising technology in the interest of risk management for supply chains. Other rising technologies are AI and IoT. There are 3 concerns that can be used to comparatively analyze these technologies (Nehme et al. 2021). First is data integrity and transparency. As blockchain's premiere feature is its immutability, blockchain trumps the other two technologies. AI and IoT can also be built to be unchanging, however, it is not intrinsically so. This extra investment and development needed for AI and IoT can be too much for some adopters.

Next is integration. Integrating new technologies with existing ones can be quite challenging. Comparing between these 3 technologies, AI and IoT provide easier integration with existing systems. Reasons for easier integration of the two technologies are the established ecosystems as well as infrastructure compatibility that is the result of these technologies being in use for longer than blockchain (Pillai et al. 2022). Lastly is the scalability. Most times, the reason organizations change their technology is to meet market standards or exceed them to become more efficient and scale easier. Scalability is a top concern. There seem to be no favorites in this category as all three technologies have their own bottlenecks. Blockchain has issues with its limited number of transactions per second, which can hinder accelerated scaling. IoT and AI face challenges if their hardware cannot keep up with the amount of work being done, leading to extra costs for infrastructure when scaling (Mirvis et al. 1991).

8.13 Discussion

The implementation of blockchain technology appears to hold some of the solutions that the supply chains of all industries have long been searching for. Main issues within supply chains are encapsulated within 4 risk categories: operational, financial, compliance, and strategic risk. The solutions offered by this technology come from its unique decentralized ledger structure, allowing the increase of transparency, accountability, security, and efficiency. Blockchain technology has multiple mechanisms by which it can add value to the supply chain by mitigating risk. Enhancing transparency, one of its core strengths, can be done via real-time tracking. By tracking every stage of a supply chain, different transactions and movements of goods, it creates an unchangeable ledger which can later be referred back to later on in the process. Doing this can reduce fraud and counterfeiting, since there is an unaltered record for every transaction, the risk of counterfeits drastically decreases. This can be best used in industries such as luxury goods, where the products are valuable and there is a tendency for bad actors to replicate these goods and sell them as genuine products. The amount of economic activity occurring with counterfeit products is not small, as of 2016 it accounted for nearly 3.5% of international trade. The reduction of the uncertainty that consumers have to face when purchasing these items could be a main driver of the adoption of blockchain technology for companies in the industries listed in Fig. 8.1.

Blockchain technology also has mechanisms in order to increase efficiency. Smart contracts are a feature of blockchain technology that holds budding potential. The main premise of smart contracts is as follows, smart contracts are contracts with their terms written in code. These contracts, once agreed upon automatically execute once the conditions or terms are fulfilled, effectively automating a process that would once need a high number of skilled manpower. This technology has the ability to increase reliability and increase efficiency, simultaneously decreasing breach of contract as well as operational

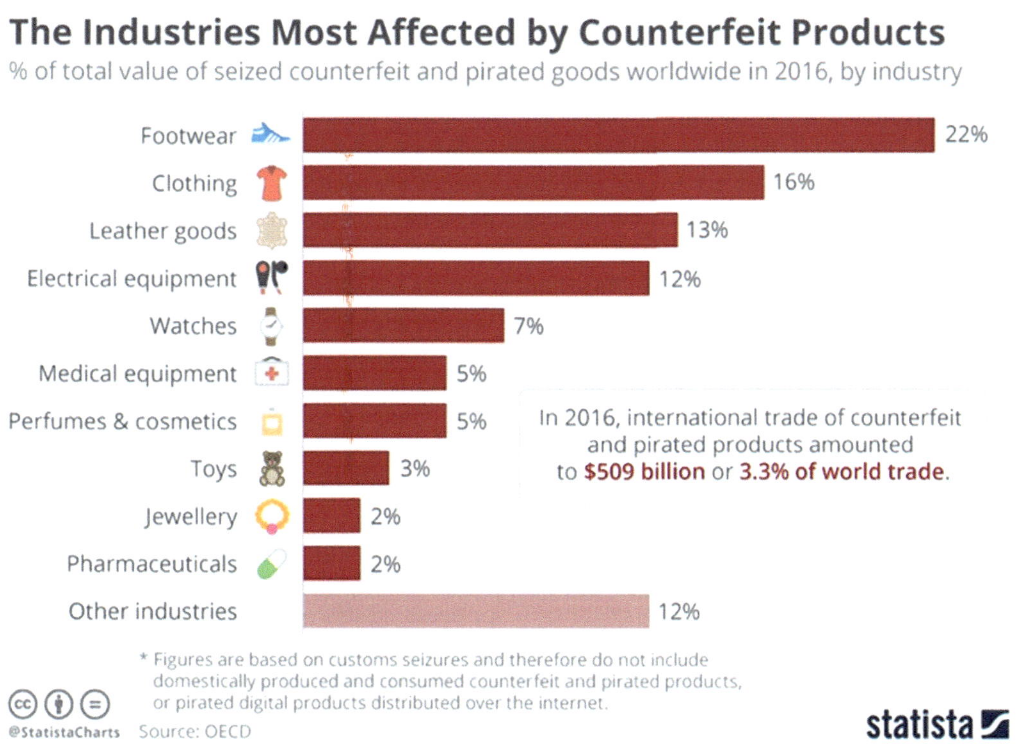

Fig. 8.1 Industries most affected by counterfeit products (Chart The Industries Most Affected by Counterfeit Products | Statista, n.d.)

costs. A case where this technology can be implemented is in the procurement and sourcing process. Using smart contracts can significantly reduce procurement time and minimize human error, resulting in a more efficient supply chain (Taherdoost 2023).

The previous two mechanisms would mostly be used in addressing operational and financial risks. Two other forms of risks can also benefit remarkably from the implementation of blockchain technology. Compliance risks can be more convoluted than the first two as they involve even more numerous stakeholders. However, blockchain technology can assist the compliance process by automating compliance checks making sure that transactions are compliant. The transparent nature of blockchain technology, one of the main reasons it was built, can also help the auditing process, simplifying the transaction track record. This specific use case can be used by pharmaceutical companies, who as seen in Fig. 8.1 also struggle with cases of counterfeiting, and are under strict regulatory control to ensure their products are up to compliance standards (Maruf et al. 2024).

Strategic risks, as previously discussed, often stem from decisions regarding suppliers. Supplier selection is incredibly important for many reasons, one of them being their responsibility for adhering to standards, both quality and ethics-wise. Blockchain technology's implementation can reduce strategic risk by enhancing supplier visibility, and increasing insight into supplier practices and performance. This visibility can enhance management's decision-making agility and accuracy, benefitting from real-time data received from transactions and movements within the supply chain. Most companies with complex and even simpler supplier networks can benefit from the added agility and responsiveness of using blockchain technology (Wong et al. 2022).

Risk aversion is the stance that numerous established companies take. After recent turbulent years, it is a widely acceptable stance, as it shows shareholders that the company is not only set on growth but sustainable growth that can weather difficult economic conditions. Their supply chain is the natural place to exert a considerable amount of effort and resources. Again, blockchain technology has real-world benefits for both the company and the consumer. The benefits offered by the supply chain can be categorized quite easily and narrow the market that would appreciate it as well. In the case of a for-profit company, if the target market is deeply concerned about issues that blockchain technology could alleviate, then it would be wise to seriously consider using this technology. In the case of a humanitarian organization, where the budget is seen more as a constraint rather than something that needs to be pushed down, then it would also be wise to implement this technology, if feasible and has a relevant number of beneficiaries.

8.14 Conclusion

Every supply chain inherently has vulnerabilities. None are immune to the risks that come with the nature of supply chains. Many breakthroughs in technology have helped with the organization and communication of different parties within the supply chain, effectively enhancing risk management in its own way. Blockchain technology which was first introduced to the world along the release of Bitcoin, could have significant upside with its widespread implementation. Blockchain technology has the potential to fill a more prominent role in supply chain risk management, offering transparency, security, improved efficiency, and improved collaboration. The technology has been used and researched quite extensively, proving that it delivers on its promises, although it does come at a cost. It is critical that the supply chain industry does not lag behind technological innovations and readily examines the prospect of this technology, thereby contributing to a future with stronger, more resilient, and transparent supply chains.

Acknowledgements This article is written to equip readers, both those familiar and unfamiliar with blockchain technology, with a top-down overview of the risks and potentials of blockchain technology in supply chain risk management.

References

Abrudan DB et al (2022) Strategic leadership, environmental uncertainty, and supply chain risk: an empirical investigation of the agribusiness industry. Agric Econ (Czech Republic) 68(5). Available at: https://doi.org/10.17221/55/2022-AGRICECON

Baah C, Acquah ISK, Ofori D (2022) Exploring the influence of supply chain collaboration on supply chain visibility, stakeholder trust, environmental and financial performances: a partial least square approach. Benchmarking 29(1). Available at: https://doi.org/10.1108/BIJ-10-2020-0519

Beck R (2018) Beyond bitcoin: the rise of blockchain world. Computer 51(2). Available at: https://doi.org/10.1109/MC.2018.1451660

Chart: The Industries Most Affected by Counterfeit Products | Statista (n.d.).: https://www.statista.com/chart/17410/counterfeit-and-pirated-products-by-category/. Accessed 24 June 2024

Domnina OL, Sakulyeva TN, Bozhko LM (2022) Supply chain risk management. Int J Process Manag Benchmarking 12(6). Available at: https://doi.org/10.1504/ijpmb.2022.125840

Francis JR, Mairose BM, Tichy EM (2021) 2020—the year the world was awakened to the importance of supply chain management. Mayo Clin Proc Innov Qual & Outcomes 5(1). Available at: https://doi.org/10.1016/j.mayocpiqo.2020.10.006

Gabuthy Y (2023) Blockchain-based dispute resolution: insights and challenges. Games 14(3). Available at: https://doi.org/10.3390/g14030034

Ghadge A et al (2021) Impact of financial risk on supply chains: a manufacturer-supplier relational perspective. Int J Prod Res 59(23). Available at: https://doi.org/10.1080/00207543.2020.1834638

Giancaspro M (2017) Is a "smart contract" really a smart idea? Insights from a legal perspective. Comput Law Secur Rev 33(6). Available at: https://doi.org/10.1016/j.clsr.2017.05.007

History of blockchain | Technology | ICAEW (n.d.) Available at: https://www.icaew.com/technical/technology/blockchain-and-cryptoassets/blockchain-articles/what-is-blockchain/history. Accessed: 12 Aug 2024

How COVID-19 impacted supply chains and what comes next | EY—Global (n.d.) Available at: https://www.ey.com/en_gl/insights/supply-chain/how-covid-19-impacted-supply-chains-and-what-comes-next. Accessed: 11 Aug 2024

Implementing New Technology (n.d.) Available at: https://hbr.org/1985/11/implementing-new-technology. Accessed: 28 June 2024

Kramer KJ, Mousavi D, Schmidt M (2022) Strategic supply chain risk management artificial intelligence and big data to support strategic supply chain risk management. ZWF Zeitschrift fuer Wirtschaftlichen Fabrikbetrieb 117(5). Available at: https://doi.org/10.1515/zwf-2022-1055

Kraude R, Narayanan S, Talluri S (2022) Evaluating the performance of supply chain risk mitigation strategies using network data envelopment analysis. Eur J Oper Res 303(3). Available at: https://doi.org/10.1016/j.ejor.2022.03.016

Kusrini E, Hanim K (2021) Analysis of compliance and supply chain security risks based on ISO 28001 in a logistic service provider in Indonesia. Int J Saf Secur Eng 11(2). Available at: https://doi.org/10.18280/ijsse.110202

Lewis A (2015) Blockchain technology explained, blockchain technologies

Liang D, Fu Y, Garg H (2024) A novel robustness PROMETHEE method by learning interactive criteria and historical information for blockchain technology-enhanced supplier selection. Expert Syst Appl 235. Available at: https://doi.org/10.1016/j.eswa.2023.121107

Maruf IR, Budiman M, Rastuti T (2024) Legal compliance for cross-border supply chain: challenge and measurements. Available at: https://doi.org/10.2991/978-2-38476-208-8_19

Mirvis PH, Sales AL, Hackett EJ (1991) The implementation and adoption of new technology in organizations: the impact on work, people, and culture. Hum Resour Manag 30(1):113–139. Available at: https://doi.org/10.1002/HRM.3930300107

Nehme E et al (2021) Ai, IoT, and blockchain: business models, ethical issues, and legal perspectives. In: Internet of things, artificial intelligence and blockchain technology. Available at: https://doi.org/10.1007/978-3-030-74150-1_4

Nimmy SF et al (2022) Explainability in supply chain operational risk management: a systematic literature review. Knowl.-Based Syst. 235. Available at: https://doi.org/10.1016/j.knosys.2021.107587

Ozdemir D et al (2022) Supply chain resilience during the COVID-19 pandemic. Technol. Soc. 68. Available at: https://doi.org/10.1016/j.techsoc.2021.101847

Pattnayak P, Patnaik S (2022) Beauty of block chain technology in supply chain management. Trends Sci 19(10). Available at: https://doi.org/10.48048/tis.2022.4175

Pillai B et al (2022) Cross-blockchain technology: integration framework and security assumptions. IEEE Access 10. Available at: https://doi.org/10.1109/ACCESS.2022.3167172

Pujawan IN, Bah AU (2022) Supply chains under COVID-19 disruptions: literature review and research agenda. Supply Chain Forum. Available at: https://doi.org/10.1080/16258312.2021.1932568

Ríos JR, Duque DFM, Gómez JCO (2019) Operational supply chain risk identification and prioritization

using the SCOR model. Ingenieria y Universidad 23(1). Available at: https://doi.org/10.11144/Javeriana.iyu23-1.oscr
Sansone G et al (2023) Blockchain for social good and stakeholder engagement: evidence from a case study. Corp Soc Responsib Environ Manag 30(5). Available at: https://doi.org/10.1002/csr.2477
Semenzin S (2023) "Blockchain for good": Exploring the notion of social good inside the blockchain scene. Big Data Soc 10(2). Available at: https://doi.org/10.1177/20539517231205479
Sinaga O et al (2019) Financial risks in supply chain management: causes and consequences. Int J Supply Chain Manag 8(2)
Taherdoost H (2022) A critical review of blockchain acceptance models—blockchain technology adoption frameworks and applications. Computers. Available at: https://doi.org/10.3390/computers11020024
Taherdoost H (2023) Smart contracts in blockchain technology: a critical review. Information (Switzerland). Available at: https://doi.org/10.3390/info14020117
Wong LW et al (2022) The role of cybersecurity and policy awareness in shifting employee compliance attitudes: building supply chain capabilities. Int J Inf Manag 66. Available at: https://doi.org/10.1016/j.ijinfomgt.2022.102520
Zhang Q et al (2021) Overview of blockchain data privacy protection. In: ACM international conference proceeding series. Available at: https://doi.org/10.1145/3475992.3476000
Zhang Y, Liu J, Sheng S (2020) Strategic orientations and participation intentions for technical standardization. Technol Anal Strat Manag 32(8). Available at: https://doi.org/10.1080/09537325.2020.1719057

Jeremy Ephraim Simarmata is a dedicated researcher at the Department of Business Management, Trisakti Institute of Transportation and Logistics in Indonesia. With a passion for exploring innovative solutions in logistics and supply chain management, Jeremy's academic focus reflects his commitment to addressing real-world challenges in these fields. His research interests span various aspects of transportation, warehousing, and distribution systems, with a particular emphasis on emerging technologies and their applications in optimizing supply chain operations. Jeremy's enthusiasm for continuous learning and his ability to blend theoretical knowledge with practical insights make him a promising young scholar in the dynamic world of logistics and supply chain research.

9 The Emerging Role of Technology in Mitigating Risk in the Humanitarian Supply Chain: Study Case Indonesian Archipelago Risk Mitigation Scheme for Food Security Due to El Nino

Achmad Yasir Baeda

Abstract

Indonesia, a collection of many islands situated between two continents and two oceans, is one of the few countries located on the path of several atmospheric-ocean oscillations that can trigger natural disasters such as floods due to prolonged rain, landslides, or even droughts that can reduce the stability of food security by changing planting and harvest schedule. Anomalies that occur in the movement of El Nino, La Nina, Madden-Julian Oscillation (MJO), Pacific Decadal Oscillation (PDO), Indian Ocean Dipole (IOD), and even Rossby Waves have significantly changed the situation and interaction between the atmosphere, the land, and the sea in Indonesia, respectively. These changes can be fatal if proper mitigation is not carried out for each anomaly. If communities, governments, and industries do not act fast, the risk of facing even more devastating future disasters will be a reality. The need for robust mitigation strategies is paramount, and all those sides must conduct quantitative and qualitative steps in substantial ways and at a timely pace since many of the anomalies have also rapidly emerged in the past two decades.

Keywords

Disaster · Mitigation · Food security · Atmospheric-Ocean Oscillation · Anomalies

9.1 Definitions

As one climate phenomenon, which basically happens every 2–7 years or more, El Niño is due to stepping up temperatures of the Pacific Ocean, changing rainfall and sea winds, currents, and most of all, sea surface temperatures. Typically, it can last a year or so, although the event will differ not every year. Declaration of an El Niño is if the Pacific SST, especially in the east, is higher than what is typical for the long term. Using the Nino 3.4 index, El Niño conditions are monitored when the threshold (the average of sea surface temperature anomalies in the 5N-5S, 120W–170W region) is greater than 0.65°, derived from a 3-month mean over a span of five or more consecutive months (Fig. 9.1a, b).

A. Y. Baeda (✉)
Marine Disaster and Weather Anomalies Research Laboratory (MDWA-RL), Faculty of Engineering, Universitas Hasanuddin, Makassar, Indonesia
e-mail: baeda@eng.unhas.ac.id

N. A. A. Rahman et al. (eds.), *Supply Chain in Humanitarian Operations*, Sustainable Development Goals Series,
https://doi.org/10.1007/978-981-96-4426-1_9

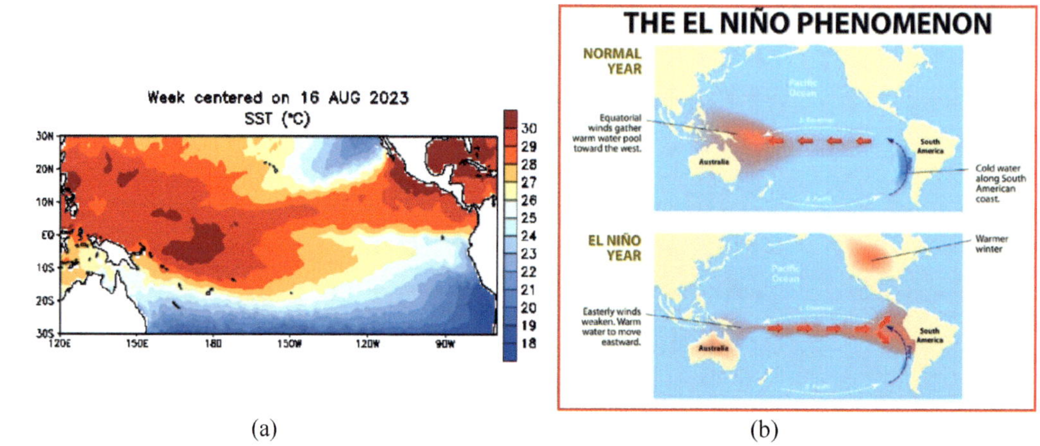

Fig. 9.1 Illustration of the movement of El Niño **a** Regarding sea surface temperature. **b** Regarding normal and El Niño year

The NOAA, on 8th June 2023, officially said that the rest of 2023 and 2024 would be an El Nino year, based on data from several years before. 2015–2016 was El Nino's last to hit the highest temperature globally, affecting around 60 million people. Around 15 million tons of rice are lost due to hotter temperatures and extended drought time. It was the same condition in Vietnam, where ground waters were over-extracted, making holes for saltwater intrusion up to 90 km, and cost around $674 million. In other Southeast Asian countries, such as The Philippines, there were affecting more than 200,000 rice and corn production, causing riots and eventually resulting in an 11% drop compared to 2014.

After the uncertainty power of 2023, El Niño predicted will make a record-breaking high temperature in 2024 due to its unknown anomaly. The dropping production rate of rice and corn will be one of the signs of it, while several experts said it might be a 20-year shortfall, more than the last event in 2015–2016.

9.2 Phases of El Niño

While being known as one of natural atmospheric circulation, corresponding to the Pacific Ocean, El Niño is a periodic changing sea surface temperature along coastal areas of Peru. The change temporarily replaces the colder current along the same coastal area.

The climate pattern that reoccurs periodically also involves changes in sea temperatures, both in the Central and Eastern parts of the Pacific; sea level pressure, winds, and rainfall; these are known as the El Niño—Southern Oscillation or ENSO (Fig. 9.2). All the Sea Surface Temperatures, or SSTs in Niño 3.4 region, are averaged-calculated monthly and then compared pre- and after-calculating months. This scheme is then used to find the deviation from the 30-year to the three-month average, which is well known as the Oceanic Niño Index or ONI.

By NOAA, ONI is then used as a primary tracker for the climate pattern of ENSO, especially regarding changes in the ocean part. The data of the three-month average temperature anomaly of surface waters, or difference from average, from the East to Central Tropical part of the Pacific is presented in Fig. 9.3. If the value is +0.5 or more, it will be said to be an El Nino, while if the value is −0.5 or lower, it will be said La Nina.

9.3 Other Atmospheric Oscillations

Indian Ocean Dipole or IOD is more like ENSO, but in the form of an irregular SST oscillation in the Indian Ocean, which is warmer in

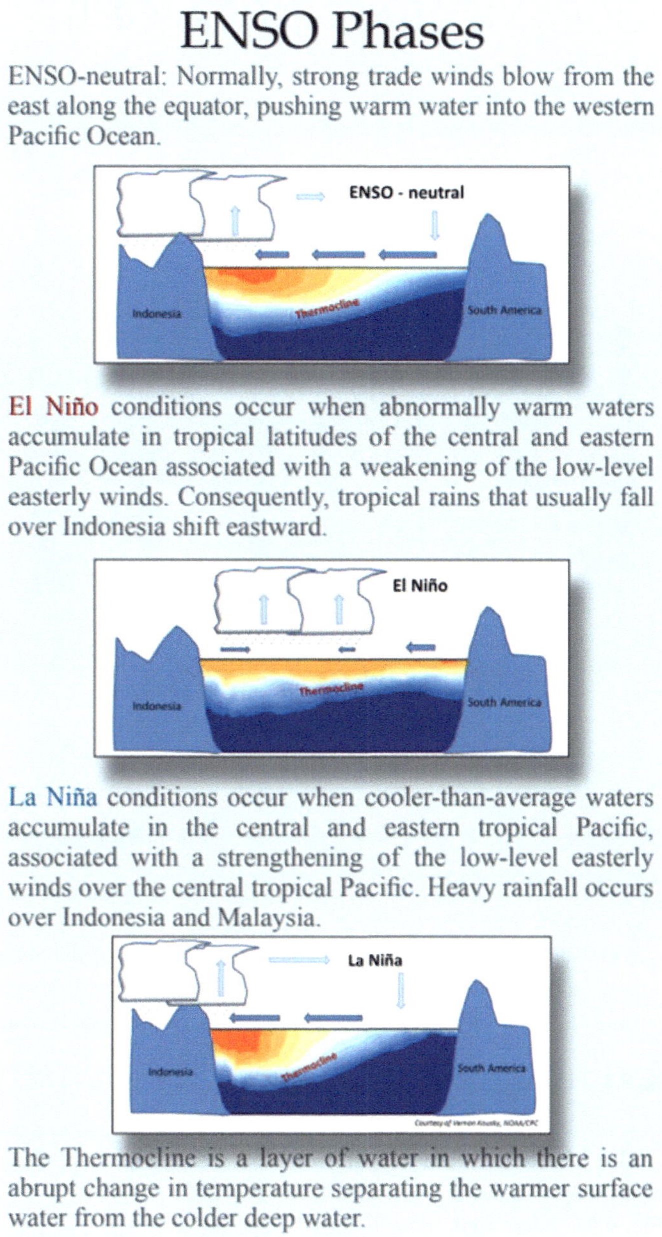

Fig. 9.2 The phases of ENSO

the western part, called the positive phase, and colder in the eastern part (called the negative phase). After June 2023, a positive phase of IOD occurred in the form of an increased sharply value (weekly value ending on 1 October 2023 was +1.69 °C) and stayed above the positive value of +0.4 °C for over seven weeks, which is an IOD event. In 2022, the negative phase of IOD occurred during the monsoon. Besides that, most of the time, it will be a neutral IOD, followed by a negative, then a positive phase, and so on. Due to this scheme, 2023 is detected as a positive IOD, followed by a negative one. The event that starts around May or June will have the highest impact in September and/or October, then recedes drastically in autumn (Fig. 9.4).

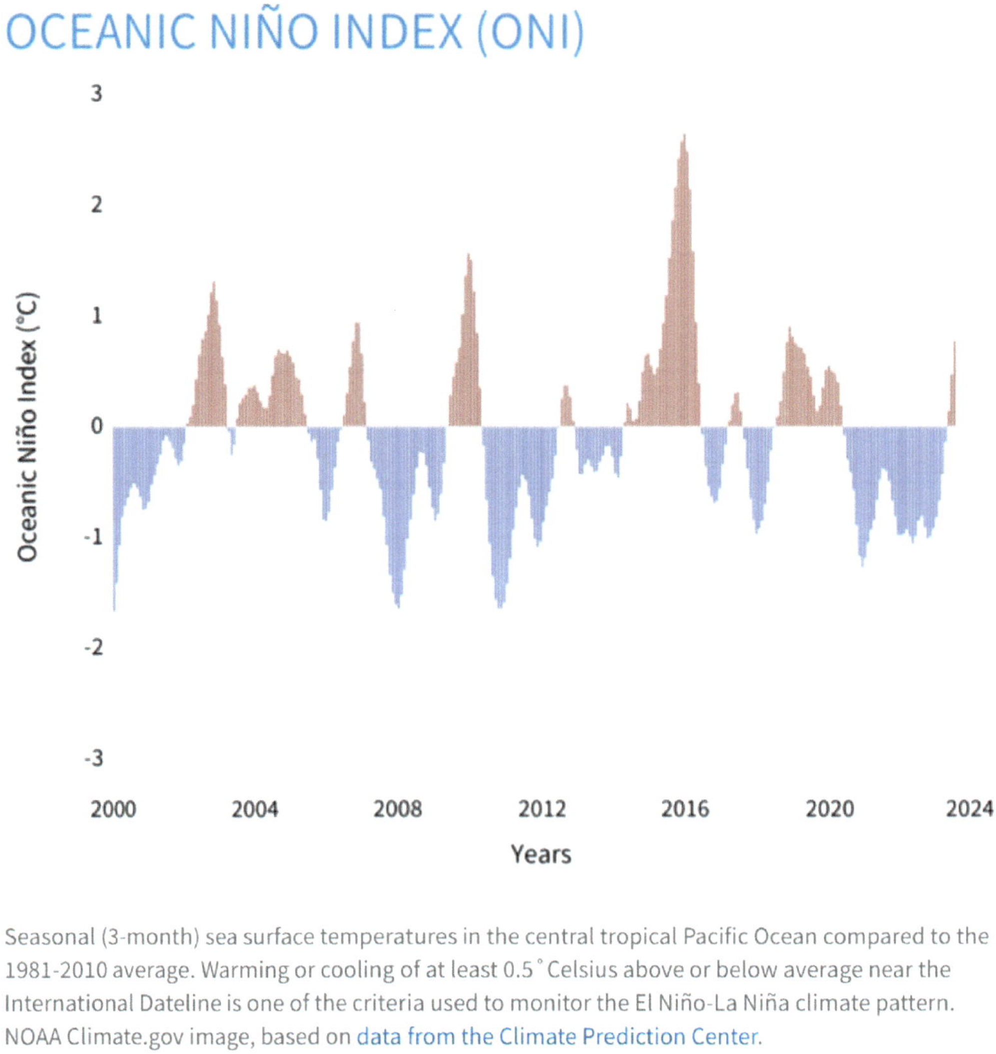

Fig. 9.3 Fluctuations of ONI in seasonal (3-months) from 2000 until 2023

The Madden-Julian Oscillation will likely weaken by the second week of October 2023 (Fig. 9.5). After that, it will slowly spread to the Western Hemisphere in Phases 8 & 1. As a result, the Western Pacific and Indian Ocean usually reduce tropical cyclone activity in these areas. The MJO is a large-scale (meso) tropical weather fluctuation, lasting weeks to months. It is generally characterized by the appearance of an eastward "pulse" of clouds and rainfall near the equator, with a 30–120-day cycle.

9.4 The Impact of El Nino

The ongoing El Niño event is expected to persist until at least April 2024. According to the latest data from the World Meteorological Organization (WMO), it will affect typical weather patterns and contribute to further high temperatures over land and mostly ocean. As of mid-October 2023, the SST with all parameters and atmospheric-oceanic indicators in the Eastern to Central tropical Pacific will be in El

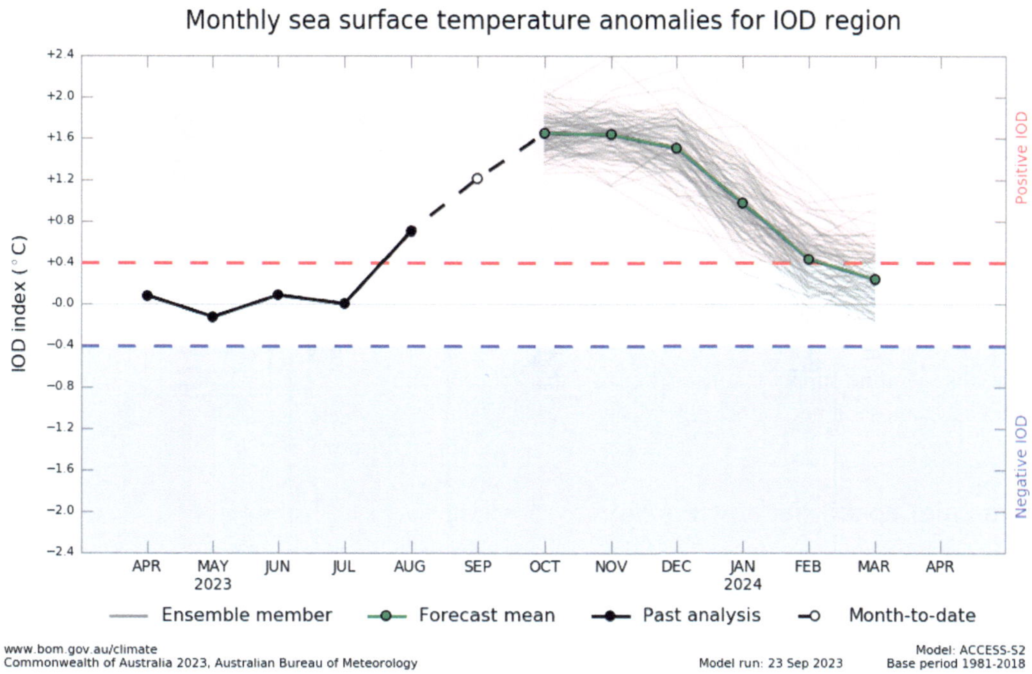

Fig. 9.4 The IOD index from April 2023 until March 2024

Niño, while ENSO will be in the warm phase. From July to August, El Niño was rapidly developed, gaining strength in September 2023 and making a decisive event in November 2023 until January 2024. Within this trend, there will be a high chance that El Niño will proceed until winter in the Northern Hemisphere or until summer in the Southern Hemisphere (Fig. 9.6).

El Niño can either positively or negatively impact ecosystems and human communities. For example, in the Southern United States, El Niño typically brings increased rainfall and sometimes damaging flooding during the fall and spring. While most commonly, La Niña will make drier weather in the Southern States, the Northwest States sometimes become warmer or even calmer and wetter than average. The most interesting thing is even though El Niño occurs in the Pacific Ocean, it can occasionally reduce the hurricane's quantity in the Atlantic Ocean. On the contrary, La Niña events tend to be one of the primary indicators for Atlantic hurricanes (Fig. 9.7).

According to all the indicators mentioned earlier, ENSO is and will tend to affect natural food production. For example, when El Niño occurs, the ocean current as one parameter will change the climate, changing marine life and affecting the income of fishermen and live consumers who depend on certain fish for food. As the ENSO can affect rainfall and altered temperature, it will change the production and availability of natural food, making agriculture highly dependent. When mostly like the Gorilla El Niño events in 1997–1998 and 2015–2016, production of rice and livestock in Southeast Asia significantly declined, causing inflation in rice prices by up to 16% globally. Based on those extreme events, several experts predict that at the end of 2023 and the beginning of 2024, El Niño will become stronger and can influence staple food production in both winter and springtime. Ultimately, this can be intensified by mandatory harvests resulting from adverse weather conditions and inadequate storage facilities.

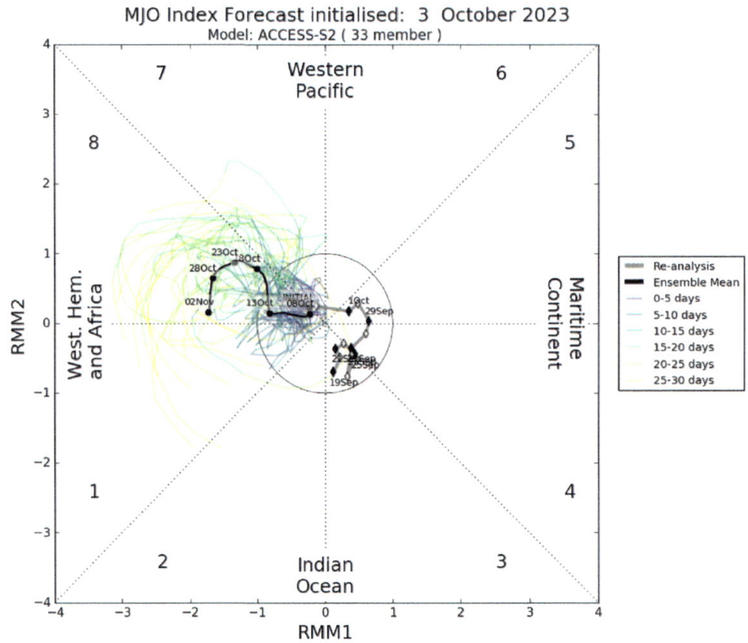

Fig. 9.5 MJO index forecast, in all 8 phases, from 19 September 2023 to 12 November 2023

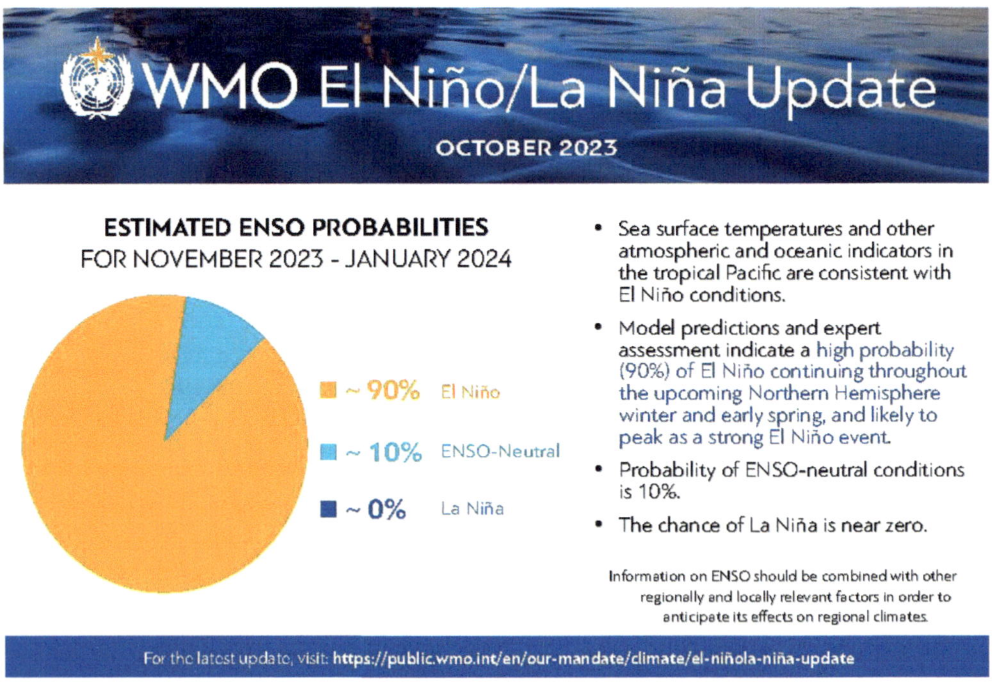

Fig. 9.6 Estimated ENSO probabilities for November 2023 until January 2024

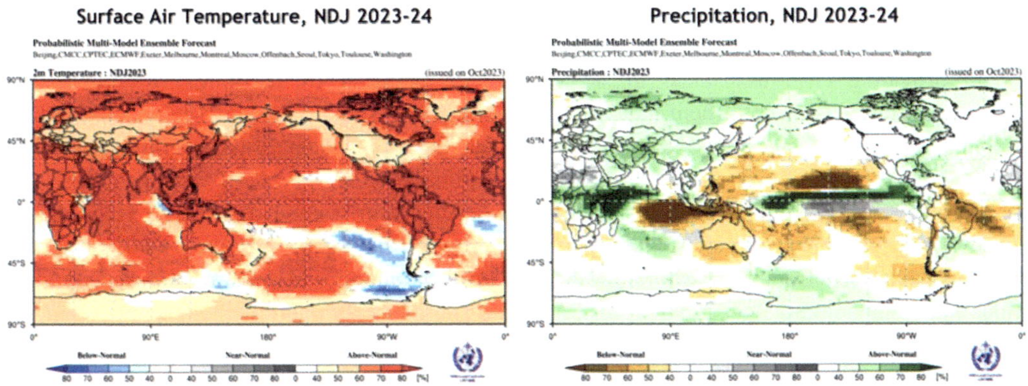

(a) (b)

Fig. 9.7 The condition of sea and land due to El Niño. **a** Effecting SST. **b** Effecting precipitations

To anticipate this, several world bodies such as FAO then published books containing general and specific information and guidelines to predict impacts that could have fatal consequences, such as the loss of one or several species of staple food crops.

El Niño-induced climate challenges pose significant threats to food security, but we have the opportunity to take proactive measures. Currently, 258 million people worldwide are experiencing acute hunger, underscoring the urgent need for increased funding; only 20% of the necessary resources for food security assistance have been secured so far. As we anticipate the potential impacts of El Niño in future times, it is essential for nations and organizations to work collaboratively to support those facing humanitarian crises. The last El Niño event in 2015–2016 affected more than 60 million individuals and led 23 countries to request international humanitarian assistance totaling USD 5 billion. By learning from these past experiences and strengthening our response strategies now, we can better safeguard vulnerable populations in future.

El Niño has the potential to significantly impact agriculture, rural livelihoods, and food security by altering rainfall and temperature patterns. Small-scale producers, including farmers, pastoralists, and fishers, are particularly vulnerable to the immediate effects of these climate shocks. Therefore, effective early action should focus on preventing damage to crops, livestock, productive lands, water resources, and infrastructure to safeguard food at its source. This approach not only secures local food supply but also lessens the broader repercussions on communities, local economies, and humanitarian aid needs. According to FAO studies, every USD 1 invested in anticipatory action can yield over USD 7 in avoided losses and additional benefits for farming families. Around USD 89 million has been called by FAO's El Niño Anticipatory Action and Response Plan to provide immediate support to the most at-risk populations in the second half of 2023.

The objectives of the Plan are three-fold:

1. To mitigate the impacts of El Niño disasters through proactive measures. This includes assisting fishers in safeguarding their boats prior to storms, reinforcing river embankments in anticipation of flooding, distributing drought-resistant seeds to rain-fed farmers, and ensuring the health of livestock.
2. To leverage the beneficial effects of El Niño while offsetting losses. For instance, supplying seeds to farmers affected by flooding enables them to plant and recover their harvests as the waters recede.
3. To implement an early response when El Niño leads to devastation. This involves

prepositioning essential supplies, such as veterinary medicines, seeds, and water bladders, and providing cash assistance to severely impacted families to help them address their immediate needs.

Rice production, like much of agriculture, is heavily influenced by climate conditions. A heat wave occurred from March to May 2023, which saw temperatures exceed 45 °C in Laos, Thailand, and Myanmar, and rise above 40 °C in Malaysia, Cambodia, and Vietnam (Bhandari 2023), led to delays in the planting time. Additionally, El Niño is anticipated to extend that condition from the middle of 2023 to the middle of 2024, causing plants to undergo overheating, which will negatively affect the development and production of rice.

Similar to humans who may suffer from heat cramps, dehydration, or heat stroke, plants also face significant adverse effects due to elevated temperatures. Research indicates that heat stress can lead to a reduction in water availability, inhibiting growth and development. This phenomenon is associated with delayed growth, diminished pollination efficacy, compromised seedling and root development, and visible symptoms such as wilting or chlorosis of leaves. Additionally, heat stress may result in a decreased number of tillers, which are crucial for grain production, and in severe cases, can culminate in the mortality of plant seedlings (WHO).

Temperatures exceeding 33 °C have been shown to significantly diminish rice yields (Arshad et al. 2017). A study conducted by the Philippines IRRI indicated for each 1 °C rise in average nighttime temperatures throughout the dry season, there is an associated average yield loss of 10% (Peng et al. 2024). With the likelihood of El Niño contributing to record-breaking temperatures, a reduction in yields is anticipated.

Overheating will be mitigated through watering time; however, the El Niño phenomenon often leads to reduced precipitation, leaving rain-fed rice farms with little relief. Even irrigated farms confront competing demands for water, which is needed for residential, industrial, and biodiversity purposes.

In addition to affecting crop yields, heat increases spoilage, a situation exacerbated by the insufficient storage infrastructure in Southeast Asia and the rise of pests and diseases that thrive in warmer conditions. This heat also hampers the productivity and work capacity of farmers and those involved in the value chain, resulting in less-than-optimal supply output.

Historically, the rice production in Southeast Asia has been significantly influenced by "very strong" El Niño events (Null 2023). The El Niño phenomenon of 1997–1998 notably diminished rice output across nearly all Southeast Asian nations, with the Philippines experiencing the most considerable decline. Nonetheless, the overall production was supported by an increase in Vietnam's rice output, leading to a modest net increase of approximately 2 million tons in rice production between the 1995–1996 and 1997–1998 periods (FAOStat 2023).

The El Niño event of 2015–2016 was regarded as significantly more severe than its predecessors. Often referred to as the "Godzilla or Gorilla El Niño" (Schiermeier 2015), it led to a drop of 15 million tons in rice production in relation to the last couple of years. The repercussions on agriculture went more just rice cultivation. El Niño in Vietnam caused the loss of more than 6,000 livestock and affected 70,000 hectares of fish farming. More than two million individuals experienced the impact, with one million facing food insecurity. The Philippines also endured drought, especially in the central and southern areas, which led to a 20% reduction in fishery productivity. Approximately 2.5 million Cambodians experienced the effects of drought, as well as losses in farmland and livestock (UNDP 2018) (Fig. 9.8).

Given the record-breaking ocean temperatures observed in 2023 (Harvey 2023), El Niño is likely to be classified as "very strong" by the end of 2023. This phenomenon often leads to warmer temperatures in North Vietnam, Thailand, Myanmar, and Laos, while causing more warm and dry conditions

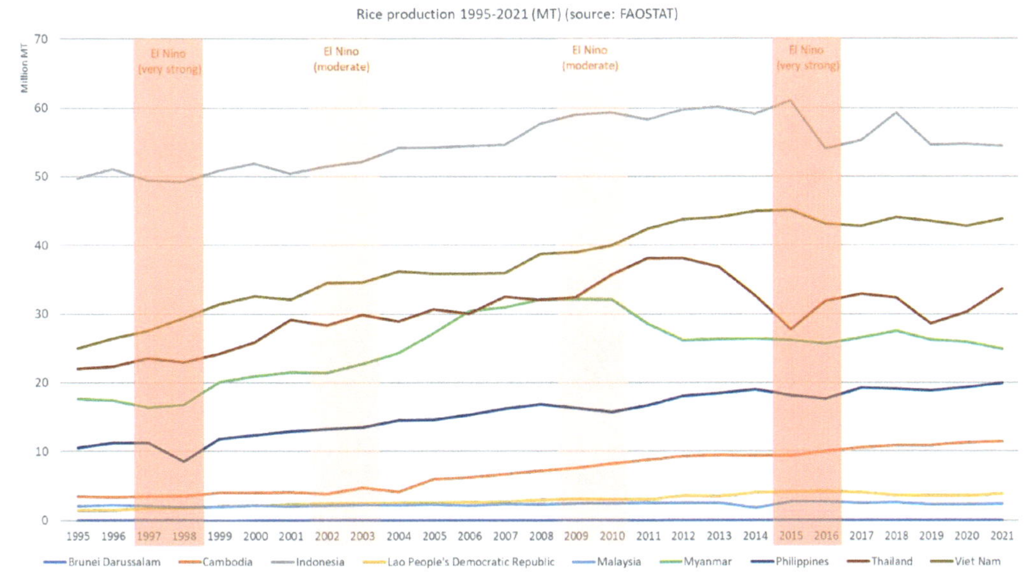

Fig. 9.8 The production of rice in Southeast Asia has been adversely affected during periods of "extreme" El Niño events, particularly in countries such as Indonesia, Thailand, Myanmar, and the Philippines (GGweather 2023)

in The Philippines, South Vietnam, Indonesia, Malaysia, and Cambodia (Fig. 9.9).

El Niño is expected to affect nearly all rice-growing regions of Southeast Asia, with a particularly pronounced impact on winter and spring rice crops.

Fortunately, El Niño in Southeast Asia nations have drawn lessons from previous ones and have made investments in readiness, prevention, and minimizing risk initiatives. These nations have established planning strategies at both national and local levels, aimed at establishing systems that are informed by risk, well-resourced, and coordinated effectively.

Attempts have been made to create rice varieties that can withstand drought, but additional efforts are still necessary, and it is improbable that many farmers will experience the advantages right away. Comprehensive irrigation and water management strategies have been implemented to tackle the anticipated drought. For instance, Thailand has created a nationwide water management plan (Wipatayotin 2023). Meanwhile, Malaysia has set up a "war room" to observe reservoir levels. Vietnam is in the process of developing a drought response plan that will be effective until 2025, anticipating the potential impacts of El Niño over the next three years (Vietnam News 2023). Using a lot of each Southeast Asian country experienced throughout several years will enhance the domestic preparedness for El Niño impact in future.

In the context of strengthening food security, several strategic measures can be implemented.

1. **Promoting Awareness Among Consumers Regarding the Diversification of Staple Foods**
   Consumers may soon face rising rice prices and potential shortages due to an anticipated shortfall. While alternatives to rice exist, many consumers are hesitant to make the switch. There is a pressing need for consumer education to help diversify reliance on rice. This effort often falls outside the typical responsibilities of agriculture ministries. It must be embraced by organizations focused on food safety and robustness, together with the media, influencers, and the private sector, creating a comprehensive, community-wide effort. Like cassava, a crop native to Southeast Asia that is resilient to drought. It

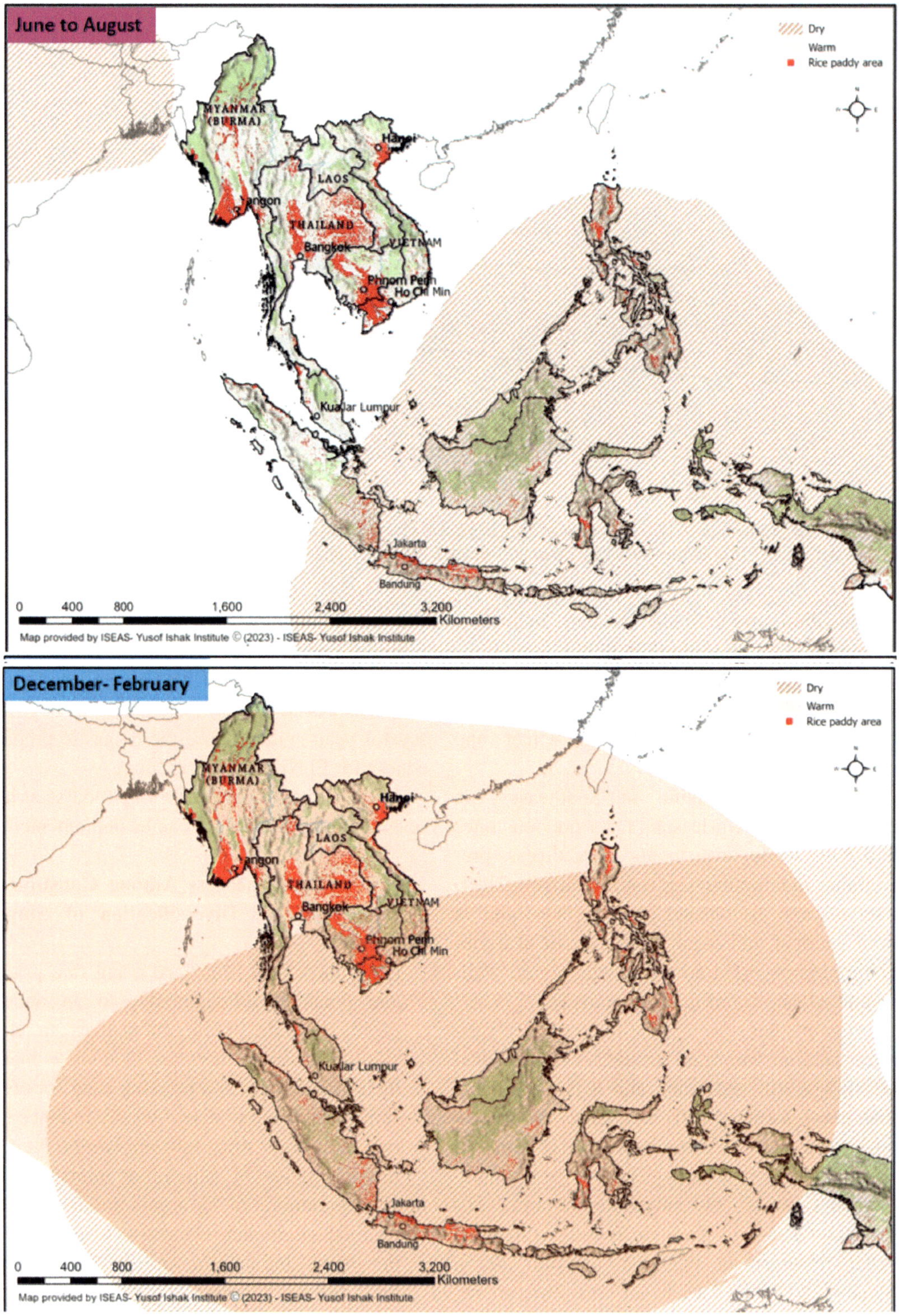

Fig. 9.9 All major rice cultivation zones are situated within regions that are likely to be impacted during the spring paddy and winter seasons. *Source* NOAA and Han et al. (2021)

has been written as a staple food, and it has come to be viewed as inferior due to pro-rice policies. Although it remains widely consumed in the area, it is considered a viable alternative to rice. Yet, its roots boast a glycemic lower index and are high in Vitamin C and fiber, as well as other nutrients. For Cassava, Indonesia, Cambodia, Vietnam, and Thailand are significant worldwide producing nations. Notably, in Indonesia, a notable company has achieved significant success in processing cassava to mimic the texture and flavor of traditional rice (Rustandi 2020). Innovations like this must be more broadly disseminated to motivate consumers to see cassava as a valid alternative.

2. **To Bolster Food Security and Ensure Sustainability in Level National and Regional**
   Many Southeast Asian countries maintain stockpiles that require enhancement or replenishment (Teng and Darvin 2019). In 2011, due to producing rice, the ASEAN Plus Three Emergency Rice Reserve (APTERR) was established successfully to stabilize the markets several times. For instance, during the crisis time of rice hyperinflation in the Philippines 2008, Japan's decision to release its reserves helped calm the markets (Timmer 2023). Nevertheless, experts widely agree that there is room for improvement in the effectiveness and responsiveness of rice stockpiles. Countries in ASEAN with a surplus of rice should be motivated to provide a larger contribution to APTERR, which currently relies primarily on contributions from the "Plus Three" countries (South Korea, China, and Japan). Furthermore, ASEAN might think about enhancing its communal stockpile database to a real-time system, enabling more effective coordination and robust times to achieve its goals, particularly as various shortages emerge.
3. **Enhancing Food Security Through Community-Based**
   Since it tends to be a gradual issue, the El Niño impact contrasts sharply with sudden disasters like extreme storm events. Consequently, while its effects are felt spanning various cities, towns, and villages, they are frequently dealt with inconsistently. Therefore, it is essential to involve additional community-focused organizations that can promptly recognize and tackle community challenges, collaborating closely with authorities and community organizations. Similar to the response during the COVID-19 pandemic, it shows an increase in nonprofit groups will emerge to support peoples in times of need.
   Communities should strive for resilience by developing initiatives focused on farming within communities and households that foster resourcefulness and mitigate the impact of potential food inflation. To support this, governments should regularly assess and document formal and informal community organizations, promoting a strong and collaborative partnership. These organizations should be encouraged to offer services and gather pertinent data and details to share with government assistance organizations, enabling more efficient aid distribution.
4. **Upholding Rice Trade**
   Through the utilization of measures like export restrictions and levies, food trade restrictions have succeeded in stabilizing local markets (Kleimann 2022). Nevertheless, these measures often have adverse effects on the procurement markets, both local and international. Trading limitations can trigger rouge, resulting in price increases and breakdowns in supply chains. This creates a chain reaction that results in cost increases for related goods, contributing to general inflation in food prices and, in the end, harming consumers across the board. Moreover, FAO Research has indicated that export limitations generally result in diminished earnings for farmers, decreased output, a drop in investment, a loss of global market position to rivals, reduced foreign exchange earnings, and damage to reputation. In 2020, in reaction to interruptions in the supply chain caused by COVID-19, ASEAN issued a statement committing to maintaining open food markets to enhance the resilience and sustainability of supply chains and promote food

security (FAO 2020). There is hope that the countries in Southeast Asia will persist in supporting open food supply chains despite rice production potentially declining.

5. **Promote Collaboration in Rice Research Between Regions and Within the Same Region**
   While most of the Southeast Asian nations navigate the challenges posed by climatic variability and food insecurity, they must cultivate good habits of collaborative partnerships and strategic management. Such initiatives are essential in addressing the multifaceted impacts of climate uncertainty, enabling these nations to collectively enhance their resilience and adaptability in the face of ongoing environmental changes.
   In addition to lessons on managing El Niño, there is a significant requirement to improve knowledge and sharing technology stemming from the development and research of rice varieties that are resilient to drought, salinity, and high temperatures. For instance, a perennial rice crop variety that was developed by China is currently being tested in several Southeast Asia countries24. This perennial rice has the capacity to produce consistent yields for up to eight harvests, thereby reducing waste, inputs, methane emissions, and labor requirements. To foster collaboration among plant breeders, researchers, biologists, agronomists, and farmers, more platforms for partnership and action are essential (Anonymous 2022).

As part of the Asia-Pacific Region, two disasters can arise for Indonesia due to prolonged El Nino: Drought and Flood. Therefore, FAO proposes pre-disaster and post-disaster activities as follows:

Priority Recommendations for Drought Management:

1. Share advance notices regarding the optimal planting period and carry out awareness initiatives and training on effective water management methods and post-harvest preservation techniques for food.
2. Supply equipment for collecting and storing water.
3. Provide financial support, such as cash-for-work programs, to enhance water storage facilities (including water points, ponds, and wells) and multi-purpose cash to help access essential needs and services, such as irrigation.
4. Supply agricultural inputs such as seeds that withstand drought, shorter growth cycles, micro-gardening tools, and micro-irrigation systems to enable quick food production during impacted seasons. Additionally, it provides drip irrigation systems for long-term crops, along with resources to help restore crop production right after a drought.
5. Disperse livestock feed before drought affects animal well-being, establish fodder reserves, and replenish small livestock and fingerlings to revive production in the drought aftermath.
6. Offer support packages for the fisheries sector.

Priority Recommendation for Floods Management:

1. Share timely alerts accompanied by valuable agricultural guidance.
2. Enhance the distribution of seed and feed storage silos.
3. Ensure the provision of animal feed at evacuation points.
4. Enhance the protection of fisheries assets by utilizing boat strengthening kits and watertight drums to ensure that fishing gear remains secure and safe.
5. Enhance support for boat repairs and initiate a program for replacing fishing gear following cyclones.
6. Encourage the distribution of root crops and strengthen the network of farms to enhance market access.
7. Provide cash assistance, including multi-purpose cash to support access to food and essential needs, and cash-for-work programs to aid in the revitalization of markets.
8. Implement a strategic plan for animal re-stocking.

References

Anonymous (2022) What is perennial rice, and why is it making history? Open Access Government, November 14. https://www.openaccessgovernment.org/perennial-rice-making-history/147157/

Arshad MS, Farooq M, Asch F, Krishna JS, Prasad PV, Siddique KH (2017) Thermal stress impacts reproductive development and grain yield in rice. Plant Physiol Biochem 115:57–72

Bhandari SR (2023) Asia's heatwaves 30 times more likely due to climate change, scientists say. Radio Free Asia, May 18. https://www.rfa.org/english/news/environment/asia-heatwave-05182023020840.html

FAOStat (2023) United Nations Food and Agriculture Organisation. https://www.fao.org/faostat/en/#home

FAO (2020) Why export restrictions should not be a response to COVID-19: learning lessons from experience with rice in Asia and the Pacific. Food and Agriculture Organisation of the United Nations. https://www.fao.org/3/ca9362en/CA9362EN.pdf

Golden Gate Weather Service (2023) El Niño and La Niña years and intensities. Golden Gate Weather Service. https://ggweather.com/enso/oni.htm

Han J, Zhang Z, Luo Y, Cao J, Zhang L, Cheng F, Zhuang H, Zhang J (2021) AsiaRiceMap10m: high-resolution annual paddy rice maps for Southeast and Northeast Asia from 2017 to 2019. Earth SystSci Data Discuss 211:1–27

Harvey F (2023) Record ocean temperatures put Earth in 'uncharted territory,' say, scientists. The Guardian, April 26. https://www.theguardian.com/environment/2023/apr/26/accelerating-oceanwarming-earth-temperatures-climate-crisis

Kleimann D (2022). Food security: the role and limits of international rules on export restrictions. Bruegel | The Brussels-based economic think tank, June 22. https://www.bruegel.org/blogpost/food-security-role-and-limits-international-rules-export-restrictions

NOAA (n.d.) El Niño theme page. NOAA Pacific Marine Environmental Laboratory. https://www.pmel.noaa.gov/elnino/impacts-of-el-nino.

Null J (2023) Note: there were only "very strong", "moderate" and "weak" El Niño events since 1995. There were no El Niño events defined as "strong". "Weak" El Niños were experienced in 2004–05, 2006–07, 2014–15 and 2018–19. Source: https://ggweather.com/enso/oni.htm

Peng S, Huang J, Sheehy JE, Laza RC, Visperas RM, Zhong X, Centeno GS, Khush GS, Cassman KG (2024) Rice yields decline with higher night temperature from global warming. Proc Natl Acad Sci 101(27):9971–9975

Rustandi C (2020) Indonesia is introducing cassava rice as a strategic shift from a popular grain. KOMPAS.com, December 16. https://go.kompas.com/read/2020/12/16/214021774/indonesiaintroduces-cassava-rice-in-a-strategic-shift-from-popular-grain?page=all

Schiermeier Q (2015) Hunting the Godzilla El Niño. Nature 526:490–491

Teng PS, Darvin BAC (2019) Food Reserves: a comparative study on food reserve management and policies in Southeast Asia. SEARCA College, Los Baños, Laguna, Philippines

Timmer P (2023) Does food protection equal food security? Lessons from ASEAN's experience with rice. FULCRUM, April 5. https://fulcrum.sg/aseanfocus/does-food-protection-equal-foodsecurity-lessons-from-aseans-experience-with-rice/

UNDP (2018) Enhancing resilience to extreme climate events: lessons from the 2015–2016 El Niño Event in Asia and the Pacific. UNDP.

Viet Nam News (2023) Response plan to El Niño is imperative, say Vietnamese experts. Asia News Network, May 19. https://asianews.network/response-plan-to-el-nino-is-imperative-sayvietnamese-experts

Wipatayotin A (2023) Farmers urged to reduce rice crops. Bangkok Post, May 11. https://www.bangkokpost.com/thailand/general/2568442/farmers-urged-to-reduce-rice-crops

WHO (n.d.) Heat and Health. World Health Organisation. https://www.who.int/news-room/factsheets/detail/climate-change-heat-and-health

Achmad Yasir Baeda is a senior lecturer at the Ocean Engineering Department, Faculty of Engineering, Universitas Hasanuddin. He also works as a researcher at the Marine Disaster and Weather Anomalies Laboratory at the Naval B Building of the Gowa Campus of Engineering Unhas. He has administrated The Ocean Technology Laboratory of Unhas since 2012, containing a large flume tank that can simulate several types of waves, including tsunamis, tropical cyclones, and freak waves. Currently, his primary research corresponds with finding suitable mitigation schemes for natural and anthropogenic disasters, like tsunamis, tropical cyclones, freak waves, extreme rain, etc.

AI-Powered Humanitarian Assistance: Exploring Opportunities and Potential Risks

10

Mohd Azizi bin Hamid,
Nadia Harnisa binti Abdul Rahman,
and Shahidah binti Ahmad Suhaimi

Abstract

Artificial Intelligence (AI) is transforming humanitarian efforts, offering new possibilities for disaster response, healthcare delivery, and resource distribution. This chapter explores the dual-edged nature of AI in humanitarian contexts, highlighting its potential to improve efficiency and outcomes while also presenting significant risks and challenges. Through real-world examples, such as AI-driven disaster response during the Nepal earthquake and refugee management in the Rohingya crisis, the chapter illustrates both the successes and pitfalls of AI integration. Key considerations include the importance of data quality, the need for continuous model refinement, and the balance between AI and human expertise. Ethical concerns, particularly regarding algorithmic bias and privacy, are also discussed. The chapter concludes with lessons learned and recommendations for future AI deployment in humanitarian operations, emphasizing the need for robust infrastructure and the careful management of AI's capabilities to ensure its benefits are fully realized while minimizing harm.

Keywords

Artificial Intelligence · Humanitarian efforts · Predictive analytics · Healthcare · Supply chain optimization

10.1 Introduction

In today's rapidly evolving technological landscape, artificial intelligence (AI) has emerged as a transformative tool across various sectors, including healthcare, finance, education, and transportation. AI's capabilities in processing vast data, identifying patterns, and making decisions are not just improving efficiency but fundamentally reshaping how these industries operate. This shift leads to more personalized services and enhanced decision-making processes, marking a significant evolution in how industries function (West and Allen 2018).

AI's impact is particularly evident in industries like healthcare, where it assists in diagnostics, personalizes treatment plans, and enhances patient care through continuous monitoring. In finance, AI improves fraud detection, optimizes risk assessments, and automates customer

M. A. b. Hamid
Centre for Academic Quality, Akademi Laut Malaysia, Melaka, Malaysia
e-mail: mohdazizi.hamid@alam.edu.my

N. H. b. A. Rahman · S. b. A. Suhaimi (✉)
Faculty of Maritime Management, Akademi Laut Malaysia, Melaka, Malaysia
e-mail: shahidah.suhaimi@alam.edu.my

N. A. A. Rahman et al. (eds.), *Supply Chain in Humanitarian Operations*, Sustainable Development Goals Series,
https://doi.org/10.1007/978-981-96-4426-1_10

interactions, illustrating its broad potential to solve complex problems and improve human lives across multiple domains. These applications demonstrate AI's ability to drive innovation and efficiency, making it a critical tool in addressing modern challenges (West and Allen 2018). AI holds significant potential in humanitarian efforts, addressing challenges like disaster response and healthcare in crisis zones. However, its deployment requires careful consideration of ethical and logistical issues to maximize benefits while minimizing risks (Boston Institute of Analytics 2024).

This chapter will explore how AI can be both a powerful ally and a potential risk in pursuing humanitarian goals, equipping readers with the knowledge and critical perspective needed to maximize AI's benefits while minimizing its risks through real-world examples, current trends, and future possibilities. The following questions have been developed for understanding AI in humanitarian efforts:

Q1: What are the main reasons for using AI in humanitarian efforts?

Q2: How can AI improve disaster response and resource distribution in humanitarian contexts?

Q3: How does AI help predict disasters and optimize resource distribution in humanitarian work?

Q4: What are the risks and challenges associated with using AI in humanitarian efforts?

10.2 Understanding AI in the Context of Humanitarian Efforts

Definition and Brief History of AI

Artificial Intelligence (AI), a branch of computer science focused on creating systems that perform tasks requiring human intelligence, such as reasoning and decision-making, has evolved significantly since its inception. The concept began in the mid-twentieth century with British mathematician Alan Turing, who laid the groundwork for machines simulating human intelligence. The term "artificial intelligence" was officially coined in 1956 by John McCarthy at the Dartmouth Conference, marking the formal beginning of AI as an academic discipline (Russell and Norvig 2020).

Early AI research focused on symbolic reasoning and problem-solving, but progress was slow due to the limitations of early computers. However, the development of machine learning algorithms in the 1980s and neural networks in the 1990s marked significant advancements, allowing computers to learn from data and improve over time (Goodfellow et al. 2016). These breakthroughs have propelled AI into a transformative force across various fields, from healthcare to finance, enabling systems to perform increasingly complex tasks.

In the twenty-first century, the combination of vast amounts of data (big data), increased computational power, and sophisticated algorithms has led to rapid advancements in AI. Today, AI technologies are used in a wide range of applications, from autonomous vehicles to personalized medicine, and are increasingly being integrated into humanitarian efforts to enhance the efficiency and effectiveness of disaster response and resource distribution.

Current Applications of AI in Humanitarian Contexts

AI is increasingly being applied in humanitarian contexts to address complex challenges related to disaster response, healthcare, and resource distribution. One of the most significant applications of AI in this field is disaster response, where AI systems analyze satellite imagery, social media data, and other sources to predict and assess the impact of natural disasters such as earthquakes, floods, and hurricanes. These AI-driven insights allow humanitarian organizations to deploy resources more effectively and respond more quickly to crises (Meier 2015).

In healthcare, AI is being used to improve the diagnosis and treatment of diseases, particularly in underserved and remote areas. AI-driven diagnostic tools can analyze medical images and patient data to provide accurate diagnoses, often outperforming human specialists in specific areas. For example, AI has been instrumental in detecting early signs of diseases such as tuberculosis and malaria, significantly improving outcomes in regions with limited access to healthcare facilities (Esteva et al. 2017).

AI is also enhancing the efficiency of resource distribution in humanitarian efforts. Machine learning algorithms can predict the demand for essential supplies, such as food, water, and medicine, and optimize supply chains to ensure these resources are delivered efficiently. During the COVID-19 pandemic, AI played a critical role in managing the distribution of personal protective equipment (PPE) and vaccines, ensuring that these vital resources reached the areas with the greatest need (Thakur and Mann 2020).

Examples of AI Tools Used in Disaster Response, Healthcare, and Resource Distribution

AI tools are increasingly being integrated into disaster response, healthcare delivery, and resource distribution within humanitarian efforts, offering innovative solutions to long-standing challenges. Among these tools, AI-powered drones have proven particularly invaluable in disaster scenarios. Equipped with high-resolution cameras and advanced sensors, these drones can assess damage, deliver medical supplies, and perform search-and-rescue missions in remote or inaccessible areas. For example, during the 2015 Nepal earthquake, AI-driven drones played a crucial role in mapping affected regions, providing essential data that guided relief operations (Meier 2015).

In the realm of healthcare, AI systems like IBM's Watson and Google's DeepMind have transformed the way diseases are diagnosed and treated. Watson, for instance, is capable of analyzing extensive medical literature and patient data, aiding doctors in diagnosing complex conditions and formulating treatment plans, particularly for diseases such as cancer. Similarly, DeepMind's AlphaFold has made groundbreaking progress in deciphering protein folding—a key process in drug discovery and the treatment of various diseases (Topol 2019).

AI is also revolutionizing supply chain management and resource distribution in humanitarian contexts. During the COVID-19 pandemic, AI algorithms were instrumental in managing the distribution of personal protective equipment (PPE) and vaccines. By analyzing data on infection rates, population density, and healthcare infrastructure, these algorithms prioritized resource allocation, ensuring that supplies reached the areas of greatest need (Thakur and Mann 2020).

10.3 Possibilities: How AI Can Enhance Humanitarian Efforts

Predictive Analytics: Use of AI in Predicting Disasters and Preparing for Emergencies

AI-driven predictive analytics, which uses data and machine learning to forecast future events, is crucial in humanitarian contexts for disaster prediction and preparedness. By analyzing vast datasets from sources like weather patterns and seismic activity, AI can predict natural disasters such as hurricanes and earthquakes, enabling proactive disaster planning (Kappes et al. 2012). A key advantage is AI's ability to process data in real time, providing actionable insights. For instance, AI has been used to predict aftershocks post-earthquake, improving planning and resource allocation (Boulton et al. 2021).

AI-driven predictive analytics have proven effective in humanitarian efforts, as seen during the 2010 Haiti earthquake, where AI models predicted aftershocks, helping relief organizations avoid further damage and allocate resources more effectively (Meier 2015). Similarly, during Hurricane Harvey in 2017, AI predictions of flood-prone areas enabled targeted evacuations and strategic resource deployment, mitigating the disaster's impact (Meier 2015). Additionally, machine learning models have successfully forecasted disease outbreaks by analyzing previous data and environmental factors, guiding public health responses and resource allocation to prevent further transmission (Rajpurkar et al. 2018).

Resource Allocation and Optimization: AI in Optimizing Supply Chains and Resource Distribution

AI significantly enhances resource allocation in humanitarian operations by quickly analyzing large datasets to predict demand, identify bottlenecks, and optimize supply chains. This capability is vital during disasters, where efficient distribution of resources like food, water, and medical supplies is crucial (Adivar et al. 2019). AI algorithms can accurately forecast resource needs based on factors like population density and disaster severity, enabling more effective allocation to areas most in need (Kappes et al. 2012). This optimization ensures that critical resources reach those who need them most, improving overall disaster response.

During the crises, AI is also making a significant impact in real-time decision. AI-driven platforms can integrate data from various sources, such as logistics networks, communication systems, and on-the-ground reports, to provide a real-time overview of the situation. This information can be used to adjust strategies on the fly, ensuring that resources are allocated in the most effective manner possible (Meier 2015). For example, during the COVID-19 pandemic, AI models were used to manage the distribution of PPE and vaccines, ensuring that these critical supplies were allocated efficiently to areas with the greatest need. This approach not only improved the efficiency of resource distribution but also helped save lives by ensuring that critical supplies were delivered to where they were needed most (Thakur and Mann 2020).

Data Collection and Analysis: Using AI for Large-Scale Data Processing in Humanitarian Contexts

AI enhances data collection and analysis in humanitarian operations by processing large-scale data to generate actionable insights. During disasters, AI-driven tools monitor social media and online platforms, analyzing vast amounts of data to identify trends, assess needs, and track misinformation (Meier 2015). This capability ensures that humanitarian responses are better informed, directing resources to the areas where they are most needed. By quickly and accurately analyzing data from sources like social media posts and news articles, AI plays a critical role in improving the effectiveness of disaster response efforts (Vieweg et al. 2010).

AI's have the ability to process and analyze large-scale data is particularly valuable in improving situational awareness during disasters. By providing real-time insights into the situation on the ground, AI can help humanitarian organizations make more informed decisions and respond more effectively to the needs of affected populations. This capability is especially important in complex and rapidly evolving situations, where timely and accurate information is critical to the success of humanitarian efforts (Meier 2015).

Furthermore, AI can help organizations identify trends and patterns in large datasets that would be difficult or impossible for humans to detect. For example, AI can analyze data from past disasters to identify common factors that contribute to the success or failure of humanitarian operations. This information can be used to improve future responses, ensuring that humanitarian efforts are as effective as possible.

Healthcare and Medical Assistance: AI in Diagnosing and Treating Diseases in Underserved Areas

AI is transforming healthcare delivery in underserved areas by enhancing the diagnosis and treatment of diseases where access to healthcare professionals is limited. AI-powered diagnostic tools can analyze medical images and patient data, often outperforming human specialists in specific tasks (Topol 2019). For instance, AI-driven tools have successfully detected diseases like malaria and tuberculosis in remote regions, facilitating timely treatment and reducing disease spread. These tools are crucial in areas with scarce healthcare resources, helping bridge the gap between medical service demand and the availability of trained professionals (Rajpurkar et al. 2018).

Beyond diagnostics, AI is also being integrated into telemedicine platforms that enable healthcare providers to remotely monitor and treat patients in underserved regions. These platforms analyze patient data, offer personalized treatment recommendations, and support virtual consultations. During the COVID-19 pandemic, AI-powered telemedicine was pivotal in monitoring quarantined patients, providing timely interventions, and easing the burden on healthcare facilities (Topol 2019). Additionally, mobile health units equipped with AI-driven tools like ultrasound machines bring critical medical services directly to disaster-stricken or remote areas, improving patient outcomes and reducing the need for long-distance travel for care (Rajpurkar et al. 2018).

Communication and Coordination: Enhancing Communication Between Humanitarian Agencies Through AI

Effective communication and coordination are crucial in humanitarian operations, where the ability to share information across multiple organizations can determine the success of a response. AI plays a significant role in enhancing these aspects by powering communication platforms that analyze and integrate data from sources like emails, social media, and reports, providing a comprehensive view of ongoing operations. These AI-driven platforms enable real-time communication between teams, ensuring that all stakeholders are aligned, reducing the risk of miscommunication, and improving the overall efficiency of humanitarian efforts (Madhavan et al. 2020; Meier 2015).

Moreover, AI is instrumental in facilitating better coordination of resources in humanitarian contexts. By integrating data from various sources, AI platforms offer a real-time overview of resource availability, operational progress, and the needs of affected populations. This allows for more effective resource allocation, minimizing duplication, and ensuring aid reaches those who need it most (Adivar et al. 2019). AI's ability to enhance communication and coordination thus significantly improves the efficiency and effectiveness of humanitarian operations, ultimately saving more lives and reducing the impact of disasters and crises.

10.4 Hazards: Potential Risks and Challenges of AI in Humanitarian Efforts

Ethical Considerations

Artificial Intelligence (AI) in humanitarian efforts presents significant ethical challenges, particularly in decision-making and privacy. AI algorithms, often designed to optimize decision-making processes, can inadvertently introduce bias. These biases stem from the data used to train AI systems, which may reflect existing societal prejudices or incomplete information, leading to skewed outcomes that could adversely impact vulnerable populations. For instance, an AI system designed to prioritize aid distribution might unintentionally favor certain groups over others due to biased data inputs, exacerbating inequalities rather than alleviating them (Celis et al. 2019). Moreover, the black-box nature of many AI algorithms complicates the transparency and accountability of decision-making, raising ethical concerns about the fairness and justice of AI-driven decisions.

Privacy concerns are paramount in humanitarian contexts, where AI-driven data collection and analysis pose significant risks. Humanitarian organizations often handle sensitive biometric and personal data, making stringent data protection essential. Gazi (2020) emphasizes the need for compliance with GDPR to prevent misuse of this data, as non-compliance can severely endanger data subjects. Vannini et al. (2020) further highlight the risks of improper data handling, particularly for vulnerable populations like undocumented migrants, where mishandling could exacerbate their vulnerabilities. Adhering to strict ethical standards in AI usage is crucial to protecting the rights and dignity of those in crisis situations.

Dependence on Technology

The growing dependence on AI systems in humanitarian efforts brings about significant risks, particularly in scenarios where over-reliance on these technologies could lead to disastrous outcomes. One of the primary concerns is the risk of over-reliance on AI, which might diminish the role of human judgment in critical decision-making processes. Humanitarian operations often occur in unpredictable and rapidly changing environments where nuanced understanding and adaptability are crucial. Jarrahi (2018) discusses the complementarity between AI and human decision-making, emphasizing that while AI can handle complexity with greater computational power, it lacks the holistic, intuitive approach that humans bring, especially in uncertain and equivocal situations. This gap can lead to AI decisions that may not align with the nuanced realities on the ground.

Moreover, the failure of AI systems during critical moments poses a severe risk. In a humanitarian context, the stakes are incredibly high, and the failure of an AI system—whether due to technical glitches, incorrect data, or unforeseen variables—could result in dire consequences, such as the misallocation of resources or delayed response times. These failures underscore the importance of maintaining a balance between AI and human intervention, ensuring that AI systems are used as tools to aid, rather than replace, human decision-making (Coppi et al. 2021).

Accessibility and Inclusivity

The implementation of AI in humanitarian efforts is fraught with challenges, particularly in resource-poor settings where access to technology and infrastructure is limited. In such environments, the introduction of AI technologies can exacerbate existing inequalities, creating a digital divide that leaves marginalized populations further behind. For instance, the deployment of AI in remote or underdeveloped regions may be hindered by inadequate internet connectivity, lack of access to reliable electricity, and the absence of technical expertise required to operate and maintain AI systems (Gwagwa et al. 2020). These barriers highlight the need for a more inclusive approach to AI implementation, ensuring that the benefits of AI are accessible to all, regardless of geographic or socioeconomic status.

Ensuring equitable access to AI technologies in humanitarian contexts is also a significant challenge. There is a risk that the deployment of AI might prioritize regions or groups with better infrastructure and more resources, while those in the most need are neglected. This inequity can be particularly pronounced in emergency situations where AI-driven decision-making tools are used to allocate resources or prioritize aid distribution. To address these challenges, it is crucial to develop AI systems that are adaptable to different contexts and that consider the unique needs and limitations of resource-poor settings. Barmer et al. (2021) emphasize the importance of developing human-centered AI that can adapt to different contexts and evolving user needs. This adaptability is particularly critical in environments with resource constraints, ensuring that AI systems remain effective and trustworthy.

Security Risks

The integration of AI into humanitarian systems introduces new security risks, particularly in the form of cybersecurity threats. AI-driven systems are increasingly becoming targets for cyberattacks, which can compromise the integrity and effectiveness of humanitarian operations. For instance, hackers could manipulate AI algorithms to alter the outcomes of aid distribution, leading to misallocated resources or delayed responses, which could have severe consequences for affected populations (Blauth et al. 2022). The reliance on AI in critical areas such as supply chain management and resource allocation necessitates robust cybersecurity measures to protect these systems from malicious attacks.

In addition to cybersecurity threats, there is also the risk of AI being used maliciously in conflict zones. In these environments, AI technologies could be repurposed or weaponized, potentially causing harm to civilians or exacerbating conflicts. For example, AI systems designed for surveillance or data analysis could be used by hostile actors to identify and target vulnerable populations, undermining the very humanitarian principles they were intended to support (Coppi et al. 2021). These risks highlight the need for stringent controls and international cooperation to ensure that AI technologies are not misused in ways that could endanger.

10.5 Case Studies: Successes and Failures of AI in Humanitarian Efforts

Artificial intelligence (AI) is transforming the humanitarian sector by enhancing the efficiency and scope of operations, particularly in data analysis, pattern recognition, and predictive modeling. These capabilities are revolutionizing resource distribution and decision-making in disaster-affected areas. However, the application of AI comes with challenges, as illustrated by various case studies in this chapter, which offer valuable lessons for optimizing AI's use in future humanitarian interventions.

Detailed Analysis of Specific Instances Where AI Has Been Applied

AI in Disaster Response: The Case of the Nepal Earthquake (2015)

In the aftermath of the 2015 Nepal earthquake, AI played a crucial role in disaster response efforts. Machine learning algorithms were employed to analyze satellite imagery and social media data to assess the extent of the damage and identify the most affected areas (Ghimire et al. 2022). One notable application was the use of AI-driven tools by the United Nations and various NGOs to prioritize the distribution of aid (Meier 2015). The success of this initiative demonstrated AI's potential in rapidly processing large volumes of data to support decision-making in emergency situations. The AI-driven approach significantly reduced the time required to assess the situation, allowing aid to reach affected areas more quickly. However, challenges such as the need for high-quality data and the integration of AI tools with existing systems were also highlighted.

AI Refugee Management: The Rohingya Crisis

The Rohingya crisis in Myanmar represents one of the most significant humanitarian challenges of the twenty-first century. AI was deployed to manage and streamline the registration of refugees crossing into Thailand. The United Nations High Commissioner for Refugees (UNHCR) utilized AI-driven biometric systems to register over 110,000 refugees, ensuring that aid could be distributed more effectively (UNHCR 2015).

The registration process, which was previously done manually and was prone to errors, was completed quickly and accurately, demonstrating the effectiveness of this AI application. But issues with data privacy and the possible exploitation of biometric information surfaced, problems were raised concerning the data's potential usage for uses other than humanitarian relief, such as surveillance, which raised moral problems regarding the application of AI in these situations (Beduschi 2022).

AI in Supply Chain Optimization: The Success Story of the World Food Programme

The World Food Programme (WFP) has successfully integrated AI into its supply chain operations to optimize the delivery of food aid. AI algorithms were used to forecast demand, optimize inventory management, and route planning. The implementation of these AI-driven solutions led to a significant reduction in delivery times and costs, ensuring that aid reached those in need more efficiently (Tzachor 2020; WFP 2019).

Outcome: The WFP's success illustrates how AI can enhance the efficiency of humanitarian supply chains, provided there is a robust infrastructure to support its implementation.

AI in Disease Outbreak Prediction: Mixed Outcomes in Ebola Response

During the Ebola outbreak in West Africa, AI was employed to predict the spread of the disease and allocate medical resources. AI models analyzed various data sources, including travel patterns and social media activity, to predict the trajectory of the outbreak (Bhattamisra et al. 2023). While AI provided valuable insights, the predictions were often overly optimistic or pessimistic, leading to either complacency or unnecessary panic.

Outcome: The mixed success of AI in disease outbreak prediction underscores the importance of continuous model refinement and the integration of AI with on-the-ground intelligence.

Lessons Learned from Both Positive and Negative Outcomes

The Importance of Data Quality and Integration

The case studies examined underscore the critical importance of data quality in determining the success or failure of AI applications in humanitarian efforts. AI systems require large volumes of accurate, up-to-date data to support decision-making, optimize workflows, and generate precise forecasts. For example, during the Rohingya crisis, predictive models failed due to insufficiently granular and outdated data, leading to inaccurate forecasts and misallocated resources, exacerbating challenges for aid organizations (Meier 2015). Conversely, the World Food Programme's success in disaster response and supply chain optimization highlights the effectiveness of AI when provided with reliable, well-integrated data, enabling more accurate forecasting and efficient resource allocation.

Lesson: Humanitarian organizations must prioritize establishing reliable data collection and management procedures, investing in systems that ensure data is precise, timely, and interconnected across platforms. Developing standards and partnerships for data sharing will enhance data quality, improving AI solutions' effectiveness in emergency situations.

Balancing AI with Human Expertise

While artificial intelligence (AI) offers significant benefits in processing and analyzing large datasets, it struggles to fully comprehend complex human behaviors and socio-political dynamics. This limitation was evident during the Rohingya refugee crisis, where AI models failed to account for political and cultural motivations behind migration, resulting in inaccurate forecasts and resource misallocation (Meier 2015). This highlights the necessity of complementing AI with human expertise. While AI excels in quantitative analysis, human experts are essential for interpreting results in the broader socio-political context, particularly in areas involving human behavior and decision-making. Integrating AI with human judgment ensures more nuanced and effective humanitarian responses.

Lesson: AI shouldn't be viewed as a substitute for human judgment, but rather as a tool to improve human capabilities. Organizations that support humanitarian causes need to find a balance between using AI's computational advantages and depending on human knowledge to comprehend larger social and political settings. To assist employees in interpreting AI-generated insights and making defensible judgments that take into account both qualitative and quantitative human elements, training programs ought to be created.

The Need for Continuous Model Refinement

AI models must be continuously updated and refined to remain accurate and useful, as demonstrated during the Ebola outbreak. Initially, AI models helped forecast the disease's spread and allocate resources, but as conditions changed and new data emerged, it became clear that the models needed adjustments. Without ongoing updates, these models risked becoming outdated, potentially leading to inaccurate predictions and harmful decisions. The dynamic nature of humanitarian crises means variables influencing outcomes can change rapidly, making it essential for organizations to regularly update AI models with the latest data and insights, including feedback from on-the-ground experiences.

Lesson: AI models must be continuously monitored and improved upon if they are to continue to succeed in dynamic humanitarian environments. Humanitarian organizations must set up procedures for routinely upgrading their AI models so that, as circumstances change, they continue to be accurate and applicable. This could entail forming specialized groups or alliances with the goal of improving the model and applying knowledge gained from practical applications.

Infrastructure and Scalability

The World Food Programme's success in using AI for supply chain optimization underscores the critical role of robust infrastructure and scalability in the effective deployment of AI in humanitarian operations. Implementing AI requires more than advanced algorithms; it demands a solid technological foundation that includes reliable internet connectivity, sufficient data storage, and the computational power necessary to run AI algorithms. Scalability is equally important, allowing organizations to expand these systems to meet the demands of larger or more complex operations. Without this infrastructure, even the most sophisticated AI tools may fail to realize their full potential.

Lesson: This success story highlights the need for humanitarian organizations to invest in the necessary infrastructure before integrating AI solutions. Ensuring the scalability and sustainability of AI deployments is essential, as these systems must be capable of expanding and adapting to future demands. Organizations must plan for the long-term viability of their AI initiatives, ensuring they have the resources and capacity to maintain and scale these systems as needed. This approach not only maximizes the effectiveness of AI tools but also ensures their continued impact in diverse and challenging environments.

10.6 Conclusion

The application of AI in humanitarian efforts has shown both promise and peril. While AI can significantly enhance the efficiency and effectiveness of operations, its success depends on the quality of data, the balance between AI and human expertise, continuous model refinement, and the infrastructure supporting its implementation. By learning from past successes and failures, humanitarian organizations can better prepare for the challenges and opportunities that AI presents in future.

Acknowledgements Gratitude is extended to everyone who has supported and guided throughout the completion of this book chapter. First and foremost, heartfelt appreciation goes to Assoc. Prof. Dr. Nor Aida binti Abdul Rahman for her invaluable advice and unwavering encouragement, which were instrumental in shaping the direction of this work. Sincere gratitude is also expressed to all team members for their dedication, insightful feedback, and continuous support, which greatly contributed to the development and success of this chapter. This chapter would not have been possible without the collective effort and collaboration of Prof. Aida and team members, whose contributions made this achievement a reality.

References

Adivar B, Hüseyinoğlu IÖY, Christopher M (2019) A quantitative performance management framework for assessing disaster supply chain efficiency. J Humanitarian Logistics Supply Chain

Manag 9(3):289–308. https://doi.org/10.1108/JHLSCM-04-2018-0021
Barmer H, Dzombak R, Gaston M, Palat V, Redner F, Smith C, Smith T (2021) Human-centered AI. IEEE Pervasive Comput 22:7–8. https://doi.org/10.1184/R1/16560183.V1
Beduschi A (2022) Harnessing the potential of artificial intelligence for humanitarian action: opportunities and risks. Int Rev Red Cross 104(919). https://doi.org/10.1017/S1816383122000261
Bhattamisra SK, Banerjee P, Gupta P, Mayuren J, Patra S, Candasamy M (2023) Artificial intelligence in pharmaceutical and healthcare research. In: Big data and cognitive computing, vol 7, Issue 1. https://doi.org/10.3390/bdcc7010010
Blauth TF, Gstrein OJ, Zwitter A (2022) Artificial intelligence crime: an overview of malicious use and abuse of AI. IEEE Access 10:77110–77122
Boston Institute of Analytics (2024) Exploring the future of AI: unlocking the potential. Boston Institute of Analytics. Retrieved from https://bostoninstituteofanalytics.org
Boulton CA, Shotton H, Williams HTP (2021) Using social media to monitor changes in mental health: a methodologically focused review. BMC Public Health 21(1):1–12. https://doi.org/10.1186/s12889-021-10524-8
Celis L, Keswani V, Yildiz O, Vishnoi N (2019) Fair distributions from biased samples: a maximum entropy optimization framework. ArXiv, abs/1906.02164
Coppi G, Moreno Jimenez R, Kyriazi S (2021) Explicability of humanitarian AI: a matter of principles. J Int Humanitarian Action 6(1):19
Esteva A, Kuprel B, Novoa RA, Ko J, Swetter SM, Blau HM, Thrun S (2017) Dermatologist-level classification of skin cancer with deep neural networks. Nature 542(7639):115–118. https://doi.org/10.1038/nature21056
Gazi T (2020) Data to the rescue: how humanitarian aid NGOs should collect information based on the GDPR. J Int Humanitarian Action 5(1):9
Ghimire S, Guéguen P, Giffard-Roisin S, Schorlemmer D (2022) Testing machine learning models for seismic damage prediction at a regional scale using building-damage dataset compiled after the 2015 Gorkha Nepal earthquake. Earthquake Spectra 38(4). https://doi.org/10.1177/87552930221106495
Goodfellow I, Bengio Y, Courville A (2016) Deep learning. MIT Press. http://www.deeplearningbook.org/
Gwagwa A, Kraemer-Mbula E, Rizk N, Rutenberg I, De Beer J (2020) Artificial intelligence (AI) deployments in Africa: benefits, challenges and policy dimensions. Afr J Inf Commun 26:1–28
Jarrahi MH (2018) Artificial intelligence and the future of work: human-AI symbiosis in organizational decision making. Bus Horiz 61(4):577–586
Kappes MS, Keiler M, von Elverfeldt K, Glade T (2012) Challenges of analyzing multi-hazard risk: a review. Nat Hazards 64:1925–1958. https://doi.org/10.1007/s11069-012-0294-2
Madhavan S, Bastarache L, Brown JS, Butte AJ, Dorr DA, Embi PJ, Friedman CP, Johnson KB, Moore JH, Kohane IS, Payne PR (2020) Use of electronic health records to support a public health response to the COVID-19 pandemic in the United States: a perspective from 15 academic medical centers. J Am Med Inform Assoc 27(6):943–947. https://doi.org/10.1093/jamia/ocaa060
Meier P (2015) Digital humanitarians: how big data is changing the face of humanitarian response. In: Digital humanitarians: how big data is changing the face of humanitarian response. https://doi.org/10.1201/b18023
Rajpurkar P, Irvin J, Zhu K, Yang B, Mehta H, Duan T, Ding D, Bagul A, Langlotz C, Shpanskaya K, Lungren MP (2018) CheXNet: radiologist-level pneumonia detection on chest X-rays with deep learning. arXiv preprint arXiv:1711.05225. https://arxiv.org/abs/1711.05225
Russell S, Norvig P (2020) Artificial intelligence: a modern approach, 4th ed. Pearson. https://www.pearson.com/store/p/artificial-intelligence-a-modern-approach/P100000734992
Thakur A, Mann A (2020) COVID 2019-suicides: a global psychological pandemic. Brain Behav Immun 88:952–953. https://doi.org/10.1016/j.bbi.2020.04.062
Topol EJ (2019) High-performance medicine: the convergence of human and artificial intelligence. Nat Med 25(1):44–56. https://doi.org/10.1038/s41591-018-0300-7
Tzachor D (2020) Artificial intelligence for agricultural supply chain risk management. In: CGIAR. https://cgspace.cgiar.org/handle/10568/108709
UNHCR (2015) UNHCR rolls out biometric identity management system to help Rohingya refugees in Thailand. Retrieved from https://www.unhcr.org/my/news/stories/unhcrs-new-biometrics-system-helps-verify-110000-myanmar-refugees-thailand
Vannini S, Gomez R, Newell BC (2020) "Mind the five": guidelines for data privacy and security in humanitarian work with undocumented migrants and other vulnerable populations. J Am Soc Inf Sci 71(8):927–938
Vieweg S, Hughes AL, Starbird K, Palen L (2010) Microblogging during two natural hazards events: what Twitter may contribute to situational awareness. In: Proceedings of the SIGCHI Conference on Human Factors in Computing Systems, pp 1079–1088. https://doi.org/10.1145/1753326.1753486
West DM, Allen JR (2018) How artificial intelligence is transforming the world. Brookings. Retrieved from https://www.brookings.edu/research/how-artificial-intelligence-is-transforming-the-world/
World Food Programme (WFP) (2019) Optimizing supply chains with AI: a success story. WFP Annual Report

Mohd Azizi bin Hamid is a Market Research & Publication executive at his current organization. He holds a Doctor of Philosophy from Universiti Poly-Tech Malaysia. He is also an experienced lecturer, trainer, and data analyst. He's been certified as professional Data Engineering and Certified Trainer (Halal Executive) from Halal Professionals Council Passionate to learn and improve and is experienced in managing people and the community and working together in achieving goal and also for problem-solving.

Nadia Harnisa binti Abdul Rahman is a lecturer for a maritime program at her current institution, where she is also actively using digital learning tools in her teaching and learning activities. She is a member of the Malaysian Institute of Management (MMIM) and holds a Master's degree from Universiti Teknologi MARA. Nadia Harnisa is actively involved in research publications focusing on Maritime- Services, Halal Management, Human Resource Management, Maritime Management, and Accounting. She also regularly participates in national innovation conferences to enhanced knowledge and practice in her fields of expertise.

Shahidah binti Ahmad Suhaimi is a lecturer and program coordinator for a maritime program at her current institution, where she also trains on digital learning tools. An associate member of the Malaysian Institute of Human Resource Management (MIHRM), she holds a Master's Degree from Cardiff University. Shahidah is actively involved in research publications focusing on Management, Human Resource Management, Maritime Management, and Mental Health. She also regularly participates in national innovation competitions, showcasing her commitment to advancing knowledge and practice in her fields of expertise.

11 Integration of Technology and Medical Services in Humanitarian Operations: A Case Study Approach

Nur Balqis Adriana Jailani, Nor Aida Abdul Rahman, Muhammad Danial Adam Jailani, and Suzari Abdul Rahim

Abstract

This chapter examines the pivotal role of technological integration in enhancing aeromedical services within humanitarian operations. Through a series of carefully selected case studies, we explore how emerging technologies are revolutionizing the delivery of medical care in crisis situations, particularly in remote and conflict-affected areas. The research highlights the transformative impact of innovations such as drone-assisted medical supply delivery, telemedicine applications in field hospitals, and AI-powered triage systems in aeromedical evacuations. Our analysis encompasses a diverse range of humanitarian scenarios, from natural disaster responses to complex emergencies in conflict zones. By adopting a case study approach, we provide in-depth insights into the practical challenges, successes, and lessons learned from implementing these technologies in high-stress, resource-constrained environments.

Keywords

Humanitarian operation · Medical supply · Aeromedical · Telemedicine · Field hospitals · Drone technology

N. B. A. Jailani
Manipal University College Malaysia, Melaka, Malaysia
e-mail: 241003083@scholar.manipal.edu.my
balqis.jailani06@gmail.com

N. A. A. Rahman (✉)
Universiti Kuala Lumpur, Malaysian Institute of Aviation Technology,
Subang, Selangor, Malaysia
e-mail: noraida@unikl.edu.my

M. D. A. Jailani
USM-KLE International Medical School, Belagavi, Karnataka, India
e-mail: md.adam11@student.usm.my

S. A. Rahim
Graduate School of Business, Universiti Sains Malaysia, Pulau Pinang, Malaysia
e-mail: suzari@usm.my

11.1 Introduction

Humanitarian crises, whether natural disasters or human-made conflicts, significantly impact health systems, causing various levels of disruption with mild to severe consequences for population health (Azmat et al. 2021). The frequency and intensity of humanitarian crises have increased in recent decades due to factors such as climate change, urbanization, and geopolitical instability (Grennan 2019). In this context, humanitarian medical operations play a crucial role in saving lives, alleviating suffering, and maintaining human dignity. These operations face numerous challenges, including

N. A. A. Rahman et al. (eds.), *Supply Chain in Humanitarian Operations*, Sustainable Development Goals Series,
https://doi.org/10.1007/978-981-96-4426-1_11

unpredictable demand, damaged infrastructure, limited resources, and coordination difficulties among multiple stakeholders (Syahrir et al. 2015). The complexity of modern crises requires innovative approaches to overcome these obstacles and improve health outcomes for affected populations.

The humanitarian health sector must continually adapt to address evolving challenges and meet the diverse needs of crisis-affected communities. This adaptation often involves integrating new technologies and approaches to enhance the efficiency and effectiveness of medical aid delivery. As noted by Behl and Dutta (2019), humanitarian supply chain management, particularly for medical supplies, is a critical aspect of crisis response that requires innovative solutions to overcome logistical and operational hurdles. Having noticed the importance of this topic, the authors of this chapter aim to to examine the pivotal role of technological integration in enhancing medical services within humanitarian operations, with a particular focus on aeromedical capabilities. Through a series of carefully selected case studies, we explore how emerging technologies are revolutionizing the delivery of medical care in crisis situations, particularly in remote and conflict-affected areas. The research also highlights the transformative impact of innovations such as drone-assisted medical supply delivery, telemedicine applications in field hospitals, and AI-powered triage systems in aeromedical evacuations.

11.2 The Role of Technology in Medical Services in the Humanitarian Context

In recent years, technological advancements have played an increasingly crucial role in enhancing medical services within humanitarian operations. These innovations have the potential to overcome many of the challenges faced in crisis situations, including limited resources, damaged infrastructure, and difficult access to affected populations. The integration of technology in humanitarian medical services spans various aspects of healthcare delivery, from diagnostics and treatment to supply chain management and coordination. One of the most significant technological contributions to humanitarian medical services is telemedicine. This approach allows healthcare providers to offer remote consultations, diagnoses, and treatment guidance, bridging the gap between medical experts and patients in crisis-affected areas. Telemedicine has proven particularly valuable in conflict zones and remote disaster sites where physical access for specialized medical personnel is limited or dangerous. For instance, the use of video conferencing and digital imaging technologies enables real-time consultations between field medics and specialists located in safer areas or even different countries (Kontodimopoulos et al. 2021).

In current crisis era, mobile health (mHealth) applications have also revolutionized healthcare delivery in humanitarian settings. Smartphone-based apps can provide valuable tools for health assessment, disease surveillance, and patient monitoring. These applications often incorporate data collection and analysis capabilities, allowing for real-time tracking of disease outbreaks and more efficient allocation of resources. For example, in refugee camps and other temporary settlements, mHealth tools have been used to manage vaccination campaigns, track nutritional status, and provide health education to displaced populations (Ullah et al. 2021).

Artificial Intelligence (AI) and machine learning are also increasingly being applied to improve decision-making and resource allocation in humanitarian medical operations. AI-powered triage systems can help prioritize patients based on the severity of their conditions, particularly useful in mass casualty events. Machine learning algorithms can analyze large datasets to predict disease outbreaks, optimize supply chain logistics, and identify trends in health needs across different populations (Chowdhury et al. 2021). Additionally, the use of drones and unmanned aerial vehicles (UAVs) has emerged as a game-changing technology for medical supply delivery in humanitarian contexts. Drones can overcome geographical barriers and infrastructure limitations to deliver essential medical supplies, vaccines, and blood products to hard-to-reach areas. This technology has shown particular promise in regions with limited road access or the immediate aftermath of natural disasters when traditional transportation routes are disrupted. For example, drone-based delivery systems have been successfully implemented in Rwanda and Ghana to transport blood products and medical supplies to remote health facilities (Koshta et al. 2021).

Another technology such as 3D printing technology offers innovative solutions for producing medical supplies and equipment in resource-constrained environments. In humanitarian settings, 3D printers can be used to manufacture custom prosthetics, medical instruments, and even certain types of medications. This on-demand production capability can help overcome supply chain disruptions and reduce dependence on external suppliers. Additionally, 3D-printed anatomical models can be used for surgical planning and medical training in field hospitals (Tatham et al. 2015). On the other hand, blockchain technology is being explored for its potential to improve transparency, traceability, and efficiency in humanitarian medical supply chains. By creating an immutable and decentralized record of transactions, blockchain can help prevent counterfeit medications, ensure the integrity of cold chains for vaccine distribution, and streamline the coordination of multiple aid organizations. This technology can also facilitate more efficient and transparent funding mechanisms for humanitarian health programs (Patil et al. 2021).

Other technology such as virtual and augmented reality (VR/AR) technologies are finding applications in medical training and mental health support in humanitarian contexts. VR simulations can provide realistic training scenarios for healthcare workers preparing to deploy to crisis zones, improving their readiness and effectiveness. AR applications can offer real-time guidance for complex medical procedures in field settings. In the realm of mental health, VR-based therapies are being explored as a means of providing psychological support to trauma survivors in refugee camps and conflict zones (Alam et al. 2021). While the Internet of Things (IoT) is enabling better monitoring and management of medical resources in humanitarian operations, IoT devices can track the location and condition of medical supplies, monitor the performance of medical equipment, and even provide real-time data on patient vital signs. This improved connectivity and data collection capability enhance the overall efficiency and effectiveness of medical services in challenging environments (Furstenau et al. 2022).

Despite the fact that these technological innovations offer immense potential for improving humanitarian medical services, their implementation is not without challenges. Issues of infrastructure, power supply, internet connectivity, and cybersecurity must be addressed to ensure the reliable and safe use of these technologies in crisis settings. Moreover, there are ethical considerations regarding data privacy, informed consent, and equitable access to technology-enhanced medical care that need careful consideration. The integration of these technologies also requires significant investment in training and capacity building for humanitarian healthcare workers. Adapting to new technologies and workflows can be challenging, particularly in high-stress environments. Therefore, user-friendly design and context-appropriate implementation strategies are crucial for the successful adoption of these innovations. Furthermore, it is essential to recognize that technology should complement, not replace, the human element in humanitarian medical services. The compassion, cultural sensitivity, and adaptability of healthcare workers remain irreplaceable in providing effective and empathetic care to crisis-affected populations.

As the field of humanitarian medicine continues to evolve, the role of technology in enhancing medical services is likely to grow. Future developments may include more advanced AI-driven diagnostic tools, expanded use of robotics in surgical interventions, and further integration of big data analytics for health system strengthening in fragile contexts. However, the focus should always remain on how these technologies can best serve the needs of vulnerable populations and support the work of dedicated humanitarian healthcare providers. In conclusion, technology offers transformative potential for improving the delivery of medical services in humanitarian contexts. From telemedicine and mHealth applications to drone-based supply chains and AI-driven decision support systems, these innovations are helping to overcome the unique challenges of providing healthcare in crisis situations. As we continue to develop and implement these technologies, it is crucial to ensure they are deployed ethically, sustainably, and in ways that truly enhance our ability to save lives and alleviate suffering in the world's most challenging environments.

The Role of Technology in Aeromedical Services

Aeromedical services are a vital component of humanitarian health operations, enabling rapid transport of patients, medical personnel, and critical supplies to and from disaster-affected areas. These services play a crucial role in saving lives and providing timely medical care in situations where ground transportation is impractical or impossible (VanVactor 2011; Rahman 2024, Hassan and Rahman 2023). As humanitarian crises become more complex and widespread, the demand for aeromedical capabilities continues to grow. Technology is playing an increasingly important role in enhancing aeromedical services within the humanitarian context. Advanced aircraft designs, GPS navigation systems, and real-time weather monitoring allow for safer and more efficient air operations in challenging environments. Telemedicine capabilities enable remote consultations and guidance for medical personnel in the field. Drone technology is being explored for the delivery of medical supplies to hard-to-reach areas (Koshta et al. 2021).

The integration of information and communication technologies (ICTs) is particularly transformative for aeromedical services. Kabra and Ramesh (2016) noted that ICTs can improve coordination, information sharing, and decision-making in humanitarian logistics. In the aeromedical context, this translates to better flight planning, patient tracking, resource allocation, and overall mission management. However, the adoption and implementation of new technologies in humanitarian aeromedical services is not without challenges. Issues of infrastructure, training, interoperability, and sustainability must be carefully considered (Dolinskaya et al. 2018). There is a need for evidence-based approaches to evaluate the impact and cost-effectiveness of technological interventions in this domain. As such, more study in this field is needed to provide depth understanding on the use of technology for intervention and innovation in ensuring sustainable humanitarian supply chain. The next subtopic will discuss on the three case studies of the technology used for medical services in humanitarian supply chain context.

11.3 Case Study 1: Drone-Assisted Medical Supply Delivery

Recent worldwide pandemic of the COVID-19 pandemic emphasized the importance of timely medical supply deliveries, especially in remote or underserved areas where traditional supply chains faced significant disruptions. The pandemic led to a surge in the demand for critical medical supplies, such as vaccines, medicines, and testing kits, which needed to be delivered swiftly to ensure public health safety. However, geographical barriers, limited road access, and health risks posed by human contact created challenges for conventional delivery methods. These limitations necessitated a more agile, contactless solution that could safely and efficiently transport medical supplies to isolated regions while adhering to social distancing measures (Cheema et al. 2022; Soni et al. 2024). One key challenge in delivering medical supplies was ensuring timely delivery in emergency situations. Traditional supply chains often relied on ground transport, which could be delayed due to road blockages, poor weather conditions, or logistical inefficiencies. Remote and rural areas, with limited healthcare infrastructure, were particularly vulnerable, as access to medical supplies was restricted (Soni et al. 2024). In these settings, drones emerged as a viable solution to bridge the gap by providing fast, reliable, and autonomous delivery capabilities.

Another critical challenge involved maintaining the security of deliveries. Medical supplies, especially during the pandemic, were prone to cyber-attacks and theft, which threatened both the integrity and availability of these critical resources. As drone technology evolved to address logistical issues, it became clear that without adequate cybersecurity measures, the risks of impersonation and unauthorized access to the delivery network were high (Cheema et al. 2022). The need for enhanced security protocols to protect drone deliveries was therefore paramount. Therefore, in order to address these challenges, drone-assisted delivery systems were implemented using advanced technologies, such as blockchain and machine learning, to ensure the secure and efficient delivery of medical supplies. Unmanned aerial vehicles (UAVs) equipped with these technologies became crucial in providing contactless, fast, and reliable services. Blockchain technology played an integral role in securing the delivery process. By implementing a decentralized ledger system, blockchain allowed for the registration and authentication of participating entities, such as medical centers, products, and drones (Cheema et al 2022). This system provided transparency and traceability, ensuring that only authorized drones and entities could participate in the delivery process.

Additionally, the drones themselves were equipped with GPS for navigation and IoT sensors for real-time tracking of both the drones and the supplies they carried. These sensors helped monitor temperature and humidity levels, ensuring that sensitive medical products, such as vaccines, were transported under optimal conditions (Soni et al. 2024). Furthermore, AI-driven intrusion detection systems were integrated to prevent cyber-attacks and unauthorized access during flight, adding an extra layer of security to the delivery network. The implementation of this system involved collaboration between healthcare facilities, drone manufacturers, and logistics providers. Medical supplies were packed and loaded onto drones at secure distribution hubs, and the blockchain network ensured that all entities involved were authenticated. Once airborne, drones navigated predetermined routes to deliver supplies to specific locations, such as rural clinics, with real-time tracking data shared across the network. Blockchain also ensured the integrity of these deliveries by recording all transactions on an immutable ledger, preventing any unauthorized tampering with the delivery.

On the other hand, the use of drones for medical supply deliveries during the COVID-19 pandemic yielded significant benefits. First and foremost, it drastically reduced delivery times. In several case studies, drones were able to deliver critical medical supplies within minutes, even to remote areas where traditional delivery methods would have taken several hours. This reduction in delivery time had a direct impact on patient outcomes, particularly in emergency situations where every minute counted. Furthermore, drone-assisted delivery systems contributed to minimizing human contact during the pandemic, thus reducing the risk of virus transmission. This was especially important for healthcare workers who were already at a high risk of exposure. By reducing the need for human involvement in the delivery process, drones helped to maintain social distancing measures while ensuring that medical supplies reached those in need. In addition to improving the speed and safety of deliveries, the system also enhanced the security and transparency of the supply chain. Blockchain technology ensured that all deliveries were properly authenticated and tracked, reducing the risk of theft or tampering with critical medical supplies. This enhanced security helped build trust among healthcare providers and patients, as they could be confident that their medical supplies were being delivered safely and securely.

The deployment of drone-assisted medical supply deliveries provided several key lessons for future implementations and best practices.

One of the most important takeaways was the need for a robust security infrastructure. While the use of blockchain and AI-driven security measures successfully mitigated many of the risks associated with cyber-attacks, the system still required regular updates and monitoring to stay ahead of potential threats. Additionally, the importance of securing the communication channels between drones and command centers was underscored, as any breaches in this communication could compromise the entire delivery process. Another lesson learned was the need for collaboration between multiple stakeholders. For drone deliveries to be successful, coordination between healthcare providers, logistics companies, drone operators, and technology providers was essential. These collaborations ensured that the technology was deployed effectively and that deliveries were made on time and securely.

In terms of best practices, one of the most successful strategies involved is the use of blockchain for end-to-end supply chain transparency. By implementing a decentralized ledger, all transactions related to the delivery were recorded and could not be altered, providing a high level of trust in the system. This practice not only improved security but also enhanced accountability, as all parties involved in the delivery process could be easily identified and held responsible for their roles. Lastly, the integration of IoT and real-time data analytics proved crucial for monitoring the conditions of sensitive medical supplies during transport. By utilizing sensors that provided real-time updates on temperature and other environmental factors, drones were able to ensure that products such as vaccines were delivered under the required conditions. This capability was particularly important for maintaining the efficacy of these supplies, especially in remote regions with extreme weather conditions. In conclusion, drone-assisted medical supply delivery has shown immense potential in addressing logistical and security challenges during health crises. With continued improvements in technology and stronger collaborations, drones can become a permanent fixture in medical logistics, enhancing access to critical healthcare services around the world.

11.4 Case Study 2: Telemedicine in Field Hospitals for Disaster and Humanitarian Response

The use of telemedicine in disaster response and field hospitals has gained increasing attention in recent years as a way to expand access to medical expertise in resource-constrained environments. This case study examines the potential applications and benefits of integrating telemedicine capabilities into temporary field hospitals established after major disasters, with an example—focus on earthquake response efforts. Field hospitals play a critical role in providing emergency medical care in the immediate aftermath of large-scale disasters like earthquakes. However, these temporary facilities often face significant challenges, including limited personnel, equipment, and specialist expertise to handle the surge in patients with complex trauma and other urgent medical needs (Ambroise et al. 2018). Incorporating telemedicine systems into field hospital operations can help address some of these limitations by enabling remote consultations with specialists and expanding the capabilities of on-site medical staff. The previous incident of devastating 2005 Kashmir earthquake in Pakistan provides an instructive example of both the challenges faced by field hospitals in major disasters and the potential for telemedicine to enhance emergency medical response. The 7.6 magnitude earthquake caused over 73,000 deaths and left over 128,000 people severely injured across northern Pakistan and Pakistan-administered Kashmir (Gulzari and Tarakci 2021). Local healthcare infrastructure was extensively damaged, with 574 health facilities destroyed. This created an urgent need for temporary medical facilities to treat the massive influx of injured survivors.

In the hardest-hit areas like Muzaffarabad and Balakot, field hospitals were quickly overwhelmed by the volume and severity of injuries. At one military hospital in the village of Forward Kahuta for instance, just 4 specialists and 14 generalist doctors were tasked with triaging and treating over 1,500 patients in the first 72 h after the quake (Gulzari and Tarakci 2021). The limited personnel and resources made it extremely challenging to provide timely, appropriate care to all patients, especially those with complex trauma requiring specialist intervention. While telemedicine was used on a limited scale during the 2005 earthquake response, disaster medicine experts have argued that more extensive telemedicine capabilities could have significantly enhanced the effectiveness of field hospitals and other temporary medical facilities. Dr. Azhar Rafiq, a physician involved in NASA telemedicine research, noted that with better telemedicine systems, "they could have anticipated the level of injuries en-route to the hospitals" and better prepared for incoming patients (Ambroise et al. 2018). Indeed, telemedicine offers several potential benefits for augmenting field hospital operations in disaster zones such as shown in Fig. 11.1.

In order to realize these benefits, field hospitals require appropriate telemedicine technologies and trained personnel. At minimum, this includes reliable telecommunications infrastructure, clinical video conferencing systems, and digital diagnostic devices that can transmit patient data. Equally important is having medical staff trained in telemedicine operations and protocols (Ambroise et al. 2018). Recently, the model proposed by Gulzari and Tarakci (2021) provides a useful framework for optimizing the establishment of temporary health facilities with telemedicine capabilities after major earthquakes. Their integer linear programming approach determines the optimal locations for field hospitals and allocation of medical personnel, including staff trained in telemedicine. The model considers two levels of telemedicine technology—basic "low-level" systems for routine care, and advanced "high-level" systems for complex cases requiring specialist input. Applying this model to a case study based on the 2005 Kashmir earthquake, the authors demonstrate how integrating telemedicine could have improved the disaster medical response. Their optimal solution establishes five strategically located field hospitals, with a mix of high-level and low-level telemedicine systems based

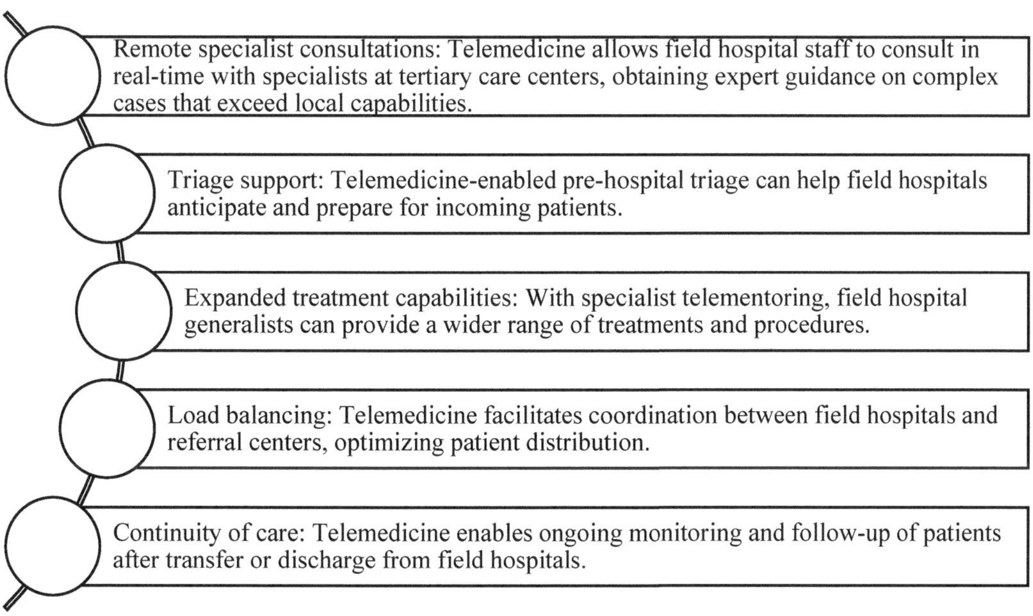

Fig. 11.1 Potential benefit for field hospital operation during disaster

on anticipated patient needs. This allows more efficient use of limited medical personnel by enabling generalist physicians to leverage telemedicine for specialist consultations.

While promising, implementing effective telemedicine in field hospitals faces several challenges. Unreliable power and communications infrastructure in disaster zones can disrupt telemedicine operations. There may also be reluctance among some clinicians to rely on telemedicine in high-stakes emergency situations. Cultural and language barriers between remote specialists and local staff/patients must be considered. Additionally, incorporating telemedicine requires additional equipment, training, and coordination that can strain already limited disaster response resources. Despite these obstacles, the potential benefits of telemedicine for enhancing field hospital capabilities warrant further exploration and pilot implementations. As telecommunications technology continues to advance and become more robust, opportunities for effective telemedicine applications in disaster response will likely expand. With careful planning and targeted investments in technology and training, telemedicine could become an integral and invaluable component of field hospital operations in future large-scale disasters.

11.5 Case Study 3: AI-Powered Triage Systems in Aeromedical Evacuations

The integration of artificial intelligence (AI) into helicopter emergency medical services (HEMS) triage systems represents an emerging frontier in emergency medical care. As HEMS continues to play a crucial role in trauma response, AI applications offer potential enhancements to patient triage, resource allocation, and clinical decision support. This case study examines the current state and future potential of AI-powered triage systems in aeromedical evacuations. A recent scoping review by Hsueh et al. (2024) highlighted the nascent but rapidly evolving role of AI in emergency medicine and HEMS. While much of the existing research has focused on hospital-based or ground emergency medical services (EMS) applications, there is growing interest in HEMS-specific AI implementations. The unique operational environment and patient population of HEMS present both challenges and opportunities for AI integration.

One of the primary applications of AI in HEMS triage is improving the identification of patients most likely to benefit from air medical transport. Reimer et al. (2022) utilized machine learning techniques to evaluate high-risk diagnostic combinations and stratify mortality risk for interfacility transport cases. Their findings suggest that AI could potentially assist in identifying cases where HEMS transfer may be futile, thus optimizing resource utilization. AI-powered triage systems may also enhance on-scene decision-making for HEMS crews. Kawai et al. (2023) demonstrated the use of a machine learning algorithm to predict the need for emergency hemostatic surgery based on HEMS-obtained vital signs and other objective data. Such predictive models could provide valuable decision support for flight crews in determining the urgency of transport and appropriate destination facilities.

The potential for AI to improve HEMS dispatch decisions is another area of active research. Atyeo et al. (2006) explored the use of intelligent decision support systems incorporating both clinical and logistical variables to assist in HEMS dispatch. Their work highlighted the potential for AI to process complex, multifactorial data to inform "go/no-go" decisions and generate optimal mission profiles.

Earlier studies on HEMS triage, such as those by Wright et al. (1989), focused on the outcomes of patients in traumatic cardiac arrest transported by helicopter. While these studies did not involve AI, they laid the groundwork for understanding the importance of appropriate triage in HEMS operations. The dismal outcomes reported in such studies underscore the potential value of AI-powered systems in improving patient selection and resource allocation. As AI technologies continue to advance, their potential applications in HEMS triage are likely to expand. Future systems may incorporate real-time physiological monitoring, advanced vital signs processing, and predictive modeling of patient outcomes. The work by Cancio et al. (2008) and Rickards et al. (2010) on processing vital signs data to predict the need for lifesaving interventions represents early steps toward such AI-powered systems.

It is important to note that while AI shows promise in enhancing HEMS triage, it should be viewed as a tool to augment, rather than replace, human clinical judgment. The complex and nuanced nature of emergency medical decision-making necessitates a human-in-the-loop approach, with AI serving as a decision support tool rather than an autonomous system. In conclusion, AI-powered triage systems represent a promising avenue for improving the efficiency and effectiveness of aeromedical evacuations. As research in this field progresses, it will be crucial to rigorously evaluate these systems in real-world HEMS operations, ensuring they enhance patient care while addressing the unique challenges of aeromedical transport. The successful integration of AI into HEMS triage has the potential to optimize resource utilization, improve patient outcomes, and advance the field of emergency medical services.

11.6 Discussion, Future Research, and Conclusion

The integration of technology in humanitarian medical services, particularly in aeromedical operations, has demonstrated significant potential to enhance the effectiveness and efficiency of crisis response efforts. Through the examination of case studies on drone-assisted medical supply delivery, telemedicine in field hospitals, and AI-powered triage systems in aeromedical evacuations, this chapter has highlighted the transformative impact of emerging technologies in overcoming the unique challenges faced in humanitarian contexts. Drone technology has proven particularly valuable in delivering critical medical supplies to remote or inaccessible areas, as evidenced by successful implementations in countries like Rwanda and Ghana. The integration of blockchain and IoT sensors has further enhanced the security and traceability of these deliveries, addressing concerns about theft and quality control. Telemedicine applications in field hospitals have shown promise in expanding access to specialist expertise in resource-constrained environments, potentially improving patient outcomes in disaster scenarios. AI-powered triage systems in aeromedical evacuations offer the potential to optimize resource allocation and support critical decision-making in high-stress situations.

However, the implementation of these technologies is not without challenges. Issues of infrastructure, power supply, internet connectivity, and cybersecurity must be addressed to ensure the reliable and safe use of these technologies in crisis settings. Moreover, there are ethical considerations regarding data privacy, informed consent, and equitable access to technology-enhanced medical care that require careful consideration. Future research in this field should focus on several key areas as outlined in Table 11.1.

As the field of humanitarian medicine continues to evolve, the role of technology in enhancing medical services is likely to grow. To conclude, future developments may include more advanced AI-driven diagnostic tools, expanded use of robotics in surgical interventions, and further integration of big data analytics for health system strengthening in fragile contexts. However, the focus should always remain on how these technologies can best serve the needs of vulnerable populations and support the work of dedicated. As such, future research and development efforts should prioritize not only technological sophistication but also adaptability, sustainability, and ethical considerations specific to humanitarian contexts.

Table 11.1 Research area for future scholars

No	Proposed future research area in the field of medical in humanitarian supply chain
1	Impact assessment: Rigorous studies evaluating the long-term outcomes and cost-effectiveness of technological interventions in humanitarian medical services are needed
2	Ethical frameworks: Development of comprehensive ethical guidelines for the use of AI and other advanced technologies in humanitarian healthcare delivery
3	Context-specific adaptations: Research on adapting technologies to diverse humanitarian contexts, considering cultural, geographical, and resource constraints
4	Integration and interoperability: Studies on improving the integration of various technological solutions to create more cohesive and efficient systems
5	Capacity building: Research on effective strategies for training humanitarian workers in the use of new technologies
6	Resilience and sustainability: Exploration of ways to make technological solutions more resilient to the challenging conditions often encountered in humanitarian operations

Source The authors

Reference

Alam F, Almaghthawi A, Katib I, Albeshri A, Mehmood R (2021) IResponse: an AI and IoT-enabled framework for autonomous COVID-19 pandemic management. Sustainability 13(7):3797. https://doi.org/10.3390/su13073797

Ambroise B, Benateau H, Prevost R, Traore H, Hauchard K, Dia H, Veyssière A (2018) The contribution of telemedicine to humanitarian surgery. J Craniomaxillofac Surg 46(8):1368–1372

Azmat SK, Ali M, Siddiqui FJ, Tirmizi SFA, Kiarie J (2021) Scoping review on the impact of outbreaks on sexual and reproductive health services: proposed frameworks for pre-, intra-, and postoutbreak situations. Biomed Res Int 9989478

Atyeo S, Sinha A, Young K (2006) A decision support system for helicopter EMS operations. In: Poster presented at: international congress of the aeronautical sciences. Hamburg, Germany

Behl A, Dutta P (2019) Humanitarian supply chain management: a thematic literature review and future directions of research. Ann Oper Res 283(1):1001–1044

Cancio LC, Batchinsky AI, Salinas J, Kuusela T, Convertino VA, Wade CE, Holcomb JB (2008) Heart-rate complexity for prediction of prehospital lifesaving interventions in trauma patients. J Trauma 65(4):813–819

Cheema MA, Ansari RI, Ashraf N, Hassan SA, Qureshi HK, Bashir AK, Politis C (2022) Blockchain-based secure delivery of medical supplies using drones. Comput Netw 204:108706

Chowdhury MJM, Ferdous MS, Biswas K, Chowdhury N, Muthukkumarasamy V (2021) COVID-19 contact tracing: challenges and future directions. IEEE Access 9:37635–37661. https://doi.org/10.1109/ACCESS.2021.3063387

Dolinskaya I, Besiou M, Guerrero-Garcia S (2018) Humanitarian medical supply chain in disaster response. J Humltarian Logist Supply Chain Manag 8(2):199–226

Furstenau LB, Sott MK, Kipper LM, Machado EL, Lopez-Robles JR, Dohan MS, Cobo MJ, Zahid A, Abbasi QH, Imran MA (2022) Link between sustainability and industry 4.0: trends, challenges and new perspectives. IEEE Access 8:140079–140096. https://doi.org/10.1109/ACCESS.2020.3012812

Grennan D (2019) What is a pandemic? JAMA 321(9):910–910

Gulzari A, Tarakci H (2021) A healthcare location-allocation model with an application of telemedicine for an earthquake response phase. Int J Disaster Risk Reduct 55:102100

Hassan A, Rahman NAA (2023) Technology application in aviation. Springer, Tourism and Hospitality

Hsueh J, Fritz C, Thomas CE, Reimer AP, Reisner AT, Schoenfeld D, Haimovich A, Thomas SH (2024) Applications of artificial intelligence in helicopter emergency medical services: a scoping review. Air Med J 43(1):90–95

Kabra G, Ramesh A (2016) Information technology, mutual trust, flexibility, agility, adaptability: understanding their linkages and impact on humanitarian supply chain management performance. Risk, Hazards & Cris Public Policy 7(2):79–103

Kawai Y, Miyazaki K, Asai H, Fukushima H (2023) Explainable prediction model of the need for emergency hemostasis using field information during physician-staffed HEMS interventions. Air Med J 42(5):336–342

Kontodimopoulos N, Angelidis P, Kostopoulou V, Niakas D (2021) Telemedicine and e-health in the Greek national health system: an overview of available services, utilization and future challenges. Int J Healthc Manag 14(3):680–687. https://doi.org/10.1080/20479700.2019.1684419

Koshta N, Devi Y, Patra S (2021) Aerial bots in the supply chain: a new ally to combat COVID-19. Technol Soc 66:101646

Patil A, Shardeo V, Dwivedi A, Madaan J, Varma N (2021) Barriers to sustainability in humanitarian medical supply chains. Sustain Prod Consum 27:1794–1807. https://doi.org/10.1016/j.spc.2021.04.022

Reimer AP, Schiltz NK, Koroukian SM (2022) High-risk diagnosis combinations in patients undergoing interhospital transfer: a retrospective observational study. BMC Emerg Med 22(1):187

Rahman NAA (2024) Urban air transport and the future of tourism. In: Contemporary marketing and consumer behaviour in sustainable tourism. Routledge, pp 28–37

Rickards CA, Ryan KL, Ludwig DA, Convertino VA (2010) Is heart period variability associated with the administration of lifesaving interventions in individual prehospital trauma patients with normal standard vital signs? Crit Care Med 38(8):1666–1673

Syahrir I, Suparno, Vanany I (2015) Healthcare and disaster supply chain: literature review and future research. Procedia Manuf 4:2–9

Soni S, Chandra P, Sharma PC, Gangrade J, Singh DK (2024) Medical kit delivery using drone: critical medical infrastructure solution for emergency medical situation. Int J Disaster Risk Reduct 108:104502

Tatham P, Loy J, Peretti U (2015) Three dimensional printing—a key tool for the humanitarian logistician? J Humltarian Logist Supply Chain Manag 5(2):188–208. https://doi.org/10.1108/JHLSCM-01-2014-0006

Ullah SMA, Islam MM, Mahmud S, Nooruddin S, Raju SM, Haque MR (2021) Scalable telehealth services to combat novel coronavirus (COVID-19) pandemic. SN Computer Science 2(1):1–8. https://doi.org/10.1007/s42979-020-00401-x

VanVactor JD (2011) Cognizant healthcare logistics management: ensuring resilience during crisis. Int J Disaster Resil Built Environ 2(3):245–255

Wright SW, Dronen SC, Combs TJ, Storer D (1989) Aeromedical transport of patients with post-traumatic cardiac arrest. Ann Emerg Med 18(7):721–726. CopyRetryClaude can make mistakes. Please double-check responses

Nur Balqis Adriana is a dedicated and passionate scholar at Manipal University College with a keen interest in exploring the intersection of medicine, logistics, and humanitarian operations. One of her interest lies in understanding how medical and dentistry skills can be applied in logistical and transportation aspects, particularly within humanitarian-based operations. She is particularly interested in the role of medical professionals in disaster relief and humanitarian aid, focusing on how their skills can be utilized to optimize logistics and ensure efficient delivery of essential medical supplies and services.

Nor Aida Abdul Rahman is an associate professor at Universiti Kuala Lumpur, Malaysia. She was a former head of Aviation Management at Universiti Kuala Lumpur, Malaysian Institute of Aviation Technology, in Subang, Selangor, Malaysia. She has been appointed as a visiting professor at ITL Trisakti Jakarta. She is also an academic fellow at UKM-MPOB Endowment Chair, National University of Malaysia. She has worked as an internal and external trainer in management, supply chain, Halal logistics, and postgraduate research. Her research work has appeared in several reputable academic journals such as Industrial Marketing Management, Journal of Islamic Marketing, and others. She has also published a number of book chapters and refereed conference proceedings, and part of the editorial team of book project with Routledge. She is a panel of WG in MS2400 Halal Supply Chain standard and TC10 for Halal Supply Chain standard (SMIIC). She earned PhD degree in Management (supply chain management) from Brunel University, London, UK. She is also serving as an academic advisor in college, a chartered member for Chartered Institute of Logistics and Transport Malaysia (CILTM), an HRDF-certified trainer, chairman (Academic Committee) for Malaysian Association of Transportation, Logistics and Supply Chain Schools (MyATLAS), vice president (Research Journal) for Institute for Research in Management and Engineering UK (INRME), JAKIM Halal-certified trainer, UniKL Halal Professional Board and a member of Academy of Marketing, UK.

Muhammad Danial Adam a passionate medical student at USM-KLE International School in Karnataka, India, is deeply committed to applying his knowledge to make a positive impact on the world. His interest in exploring the functionalities of medical in humanitarian operations stems from a desire to address global health challenges and provide essential care to underserved communities.

Suzari Abdul Rahim is a senior lecturer in Graduate School of Business, Universiti Sains Malaysia. He received his Doctoral Degree (PhD) in Supply Chain Management from Brunel University of London, United Kingdom. Prior to his academic career, he has worked for various institutions in the Manufacturing and Telecommunications industry for a number of years fulfilling Supply Chain Management roles responsible for Procurement, Purchasing, Material Management, Supplier Relationship Management, Tender Management, and Logistics. His current research interest is in the area of Halal logistics and supply chain management, investigating improvements and recommending logistics solutions for the Halal industry and community.

12 Biometric Technology in Humanitarian Sector

Nurul Ain Safwah binti Ekey Hussain, and Ramesh Babu a/l Amathalai

Abstract

Biometric technology, involving the automated identification of individuals based on unique physical or behavioral traits, has revolutionized identity verification in various sectors, including humanitarian efforts. Its adoption in refugee management, aid distribution, and healthcare delivery has enhanced the accuracy, security, and efficiency of operations in crisis situations. By integrating systems like fingerprint recognition, facial and iris scanning, and voice authentication, humanitarian organizations such as the UNHCR and the WFP have streamlined processes, minimized fraud, and ensured equitable resource distribution. However, these advancements bring challenges, including privacy concerns, data security risks, and ethical dilemmas related to informed consent and potential exclusion. As biometric technology evolves, addressing these concerns while capitalizing on its benefits will be crucial for its future use in humanitarian settings.

N. A. S. b. E. Hussain (✉) · R. B. a. Amathalai
Faculty of Maritime Management,
Akademi Laut Malaysia, Melaka, Malaysia
e-mail: Nurulain.safwah@alam.edu.my

R. B. a. Amathalai
e-mail: rameshb.amathalai@alam.edu.my

Keywords

Biometric technology · Humanitarian sector · Supply chain · Refugee management · Aid distribution · Data security · Ethical concerns

12.1 Introduction

Introduction to Biometric Technology in the Humanitarian Sector

Biometric technology, a sophisticated method of identifying individuals based on unique physiological or behavioral characteristics, has evolved significantly over centuries. Biometric technology, once a feature of science fiction, has become an integral part of modern society, particularly in sectors where identity verification and security are paramount. In the context of the humanitarian sector, this evolution marks a significant turning point, particularly in the early twenty-first century when biometric technology began to be integrated into refugee management and aid distribution. Biometrics has emerged as a vital tool in improving the efficiency and effectiveness of aid delivery, ensuring that assistance reaches those who need it most. Biometric technology has rapidly emerged as a critical tool in the humanitarian sector, offering innovative

N. A. A. Rahman et al. (eds.), *Supply Chain in Humanitarian Operations*, Sustainable Development Goals Series,
https://doi.org/10.1007/978-981-96-4426-1_12

solutions to long-standing challenges in identification, resource distribution, and data management. With the increasing complexity of humanitarian operations, particularly in conflict zones and areas affected by natural disasters, the need for accurate and efficient identification systems has become paramount.

Biometric technology refers to the automated recognition of individuals based on their physical or behavioral characteristics. This includes fingerprints, iris scans, facial recognition, voice recognition, and gait analysis (Upadhyay et al. 2020). The unique and immutable nature of these characteristics makes biometrics particularly valuable in contexts where accurate identification is crucial. These identifiers are considered highly reliable because they are difficult to replicate or forge, making biometric systems an attractive option for security and identification purposes. Table 12.1 shows the common types of biometric identifiers that are regularly used such as fingerprint, facial recognition, iris recognition, voice recognition, and gait analysis.

The adoption of biometric systems in humanitarian efforts was initially driven by the need for more reliable identification methods in environments fraught with challenges. Traditional methods, such as paper records, often proved inadequate in situations involving mass displacement due to conflicts or natural disasters. These circumstances necessitated a more robust solution, and biometric technology emerged as a reliable alternative, offering a means to accurately track and manage individuals across borders and over time.

Biometrics applications are diverse, ranging from verifying the identities of refugees and internally displaced persons (IDPs) to ensuring that aid is distributed equitably. Organizations like the United Nations High Commissioner for Refugees (UNHCR) have increasingly relied on biometric systems to manage the registration and verification of refugees in camps worldwide, thereby streamlining operations and enhancing accountability.

In this chapter, we will explore the application of biometric technology in the humanitarian sector, examining its benefits, challenges, and ethical considerations. We will also review case studies where biometric systems have been successfully implemented, discuss the impact of these technologies on humanitarian operations, and provide recommendations for their future use. The chapter will conclude with an analysis of the potential future developments in biometric technology and their implications for the humanitarian sector.

Historical Context and Evolution of Biometrics

Biometric technology, defined as the automated recognition of individuals based on their unique biological traits, has deep historical roots that trace back to ancient civilizations. The use of fingerprints as a form of identification is well-documented, with evidence of their application found in Babylonian clay seals and ancient Chinese documents (Gavrilova 2008). These early practices, though primitive by modern standards, laid

Table 12.1 The common types of biometric identifiers

Biometric Identifier	Description	Common Use Cases
Fingerprint	Recognition based on unique patterns in fingerprints	Access control, criminal identification
Facial Recognition	Analysis of facial features for identity verification	Surveillance, mobile device security
Iris Recognition	Identification through unique patterns in the iris	Border control, secure facility access
Voice Recognition	Verification of identity based on voice patterns	Call centers, mobile banking authentication
Gait Analysis	Identification based on walking patterns	Surveillance, security in public spaces

the groundwork for the development of more sophisticated biometric systems. The transition from rudimentary to scientific applications of biometrics began in the late nineteenth century, largely due to the pioneering work of Sir Francis Galton, a cousin of Charles Darwin.

Galton's research on fingerprints not only established the uniqueness of these patterns but also laid the foundation for their use in personal identification, particularly in forensic science (Mackey and Cole 2001). His work was instrumental in moving biometrics from a concept of identification to a scientific method that could be systematically applied, particularly in law enforcement. The adoption of fingerprint analysis by police forces across the world marked the beginning of biometrics as a critical tool in identity verification.

While Galton's work provided the foundation, the broader application of biometrics outside law enforcement took time to develop. It wasn't until the early twenty-first century that biometric technology began to be integrated into the humanitarian sector, a field where accurate and reliable identification is crucial. This shift was driven by the growing complexities of managing large populations affected by conflicts, natural disasters, and other crises that led to mass displacement. Traditional methods of identification, such as paper records, were increasingly proving inadequate, especially in environments where records could be easily lost, forged, or destroyed (Gelb and Clark 2013).

The early twenty-first century marked a significant turning point for the adoption of biometric technology in the humanitarian sector. The need for accurate identification systems in refugee management became particularly acute as global conflicts intensified, leading to unprecedented levels of displacement. According to the UNHCR, the global refugee population reached 25.9 million by the end of 2018, with millions more internally displaced (Altare et al. 2022). Managing such vast numbers posed significant challenges, especially in terms of ensuring that aid reached the right individuals.

Biometric systems offered a solution to these challenges by providing a means to track and manage individuals across borders and over time. The UNHCR began implementing biometric registration systems in refugee camps worldwide, starting with a fingerprint-based system in Kenya's Dadaab refugee camp in 2013. This system aimed to reduce identity fraud and ensure that aid was distributed equitably (UNHCR 2018). The success of this initiative led to the wider adoption of biometric technology in refugee management, with other organizations following suit.

The application of biometric technology in the humanitarian sector has since expanded rapidly. It is now used not only for refugee management but also for various other purposes, including the distribution of aid, health services, and even cash transfer programs. For instance, the World Food Programme (WFP) has implemented biometric systems in several countries, using fingerprint recognition to authenticate beneficiaries before they receive assistance (World Food Programme 2024). These systems have proven to be effective in reducing fraud and increasing the efficiency of aid distribution.

In conclusion, while the historical context and evolution of biometrics highlight its potential to revolutionize identification systems, particularly in the humanitarian sector, it also underscores the importance of addressing the ethical and security challenges associated with its use. As biometric technology continues to evolve, it will be crucial to develop robust frameworks that ensure its responsible application, protecting the rights and dignity of those it aims to serve.

Importance of Biometrics in Humanitarian Efforts

The adoption of biometric technology in humanitarian operations is driven by several key factors:

Efficiency

Biometrics streamline the process of identification, reducing the time and resources required to verify individuals. This is particularly important in emergency situations where speed is critical.

Accuracy
Unlike traditional methods, biometric identifiers are unique to each individual, reducing the risk of duplication and fraud.

Security
In conflict zones and other high-risk areas, biometrics can enhance security by preventing unauthorized access to resources and facilities.

Accountability
By providing a clear record of who has received aid, biometrics help ensure that resources are distributed equitably and that aid organizations are held accountable.

12.2 Applications of Biometric Technology in Humanitarian Activities

Biometric technology has been deployed in various humanitarian contexts, each with its own set of challenges and successes.

Identification and Verification in Refugee Management

One of the most significant applications of biometrics in the humanitarian sector is in the management of refugees. With millions of people displaced by conflict and natural disasters, organizations like the UNHCR have turned to biometrics to maintain accurate records and prevent identity fraud (UNHCR 2018). Accurate identification is crucial in large-scale refugee operations, where the risk of identity fraud is high, and resources are limited. Biometric systems ensure that each individual is registered once, reducing the chances of duplicate records and ensuring that aid is distributed equitably.

The UNHCR has implemented biometric registration systems in numerous refugee camps, including those in Jordan, Kenya, and Bangladesh. For example, in 2013, UNHCR implemented a biometric identity management system in Jordan's Zaatari camp, home to over 76,000 Syrian refugees. The system uses iris scans to register and verify refugees, has been a model for other refugee camps worldwide, ensuring that each refugee when they receive food and other essential supplies (UNHCR 2024b) without duplication or fraud (UNHCR 2018). The system has registered over 1.5 million refugees and the iris-scanning system has been highly effective in preventing identity fraud, a significant concern in large-scale refugee operations. By ensuring that each refugee is registered only once, the system has reduced the risk of duplicate records and ensured that resources are distributed equitably.

Cash Transfer Programs

Cash transfer programs have become an increasingly popular method of delivering aid, allowing beneficiaries to purchase what they need most. However, these programs require robust systems to verify the identities of recipients and prevent fraud. Biometric technology has proven to be an effective solution in this context, providing a secure and efficient means of verifying identities.

In Somalia, the WFP implemented a biometric system that uses fingerprint recognition to authenticate beneficiaries before they receive cash assistance. This system has significantly reduced fraud and increased the efficiency of the cash distribution process, ensuring that aid reaches those who need it most (World Food Programme 2024).

Biometric Data in Food and Aid Distribution

Biometrics have also been instrumental in food and aid distribution. The distribution of food aid is another area where biometric technology has made a significant impact. In many humanitarian operations, ensuring that food reaches the intended recipients is a major challenge. Biometric systems help to address this challenge by verifying the identities of beneficiaries at distribution points.

The WFP has integrated biometric technology into its operations to ensure that food assistance reaches the intended beneficiaries. In 2018, WFP implemented a system in Yemen, a country ravaged by conflict and food insecurity, using fingerprint scans to verify the identity of those receiving aid (World Food Programme 2024). This system has helped reduce fraud and ensure that scarce resources are distributed fairly. In Kenya's Dadaab refugee camp, the WFP uses fingerprint recognition technology to verify the identities of refugees before they receive food rations. This system has reduced the risk of fraud and ensured that food aid is distributed more efficiently (World Food Programme 2024).

In conflict-affected areas such as Somalia, the World Food Programme (WFP) has implemented biometric systems to secure food distribution centers. These systems utilize fingerprint recognition to ensure that only individuals who have been registered and verified are able to access assistance, thus preventing fraudulent claims and ensuring that aid reaches those who genuinely need it (World Food Programme 2024). The use of biometrics in such contexts not only enhances the security of operations but also helps to maintain the integrity of humanitarian programs by ensuring that resources are distributed equitably and efficiently, even in the most challenging circumstances (Gelb and Clark 2013).

Health Services

Biometric technology is also being used to improve the delivery of health services in humanitarian settings. In many cases, refugees and IDPs lack official identification, making it difficult for healthcare providers to track medical histories and provide continuity of care. By using biometric identifiers, healthcare providers can accurately link individuals to their medical records, ensuring that they receive appropriate care.

In Bangladesh, for example, the UNHCR has implemented a biometric health information system in the Rohingya refugee camps. The biometric system has been instrumental in tracking population movements within the camps and preventing fraud. By linking refugees to their biometric data, the UNHCR has been able to provide more targeted and efficient assistance, including food, shelter, and healthcare. This system uses fingerprints to link refugees to their medical records, allowing healthcare providers to track treatments and vaccinations effectively (UNHCR 2024a).

12.3 Challenges and Ethical Considerations

The adoption of biometrics in the humanitarian sector is not without its challenges. Issues related to privacy, data security, and the potential for misuse of biometric data have raised ethical concerns. The collection and storage of biometric data involve sensitive personal information, which, if compromised, could have serious consequences for individuals (Carrillo and Sequera 2020). Moreover, there is a risk that biometric systems could exclude certain individuals, particularly those with disabilities or injuries that affect their biometric identifiers (Francese et al. 2021).

While biometrics offer significant benefits, their use in the humanitarian sector is not without challenges and ethical concerns. Biometric technology offers numerous benefits in the humanitarian sector, it also raises significant ethical and privacy concerns. The collection and storage of biometric data involve sensitive personal information, which, if mishandled, could have serious consequences for individuals.

Privacy Concerns

One of the most significant concerns is privacy. Biometric data is highly sensitive and, if compromised, can have severe consequences for individuals. The collection and storage of biometric data raise questions about who has access to this information and how it is protected (Tran et al. 2021). In humanitarian contexts, where individuals are often vulnerable and displaced, the potential for misuse of biometric data is particularly concerning. There is a risk that data could be accessed by unauthorized parties, including governments or non-state actors, leading to potential harm to the individuals involved.

Data Security

One of the primary concerns with the use of biometric technology is data security. Biometric data is inherently sensitive, and its misuse can lead to serious privacy violations. For instance, if biometric data falls into the wrong hands, it could be used for unauthorized surveillance or identity theft. Therefore, it is crucial to implement robust security measures to protect biometric data from unauthorized access or breaches. Humanitarian organizations must adhere to strict data protection standards when collecting and storing biometric data. This includes ensuring that data is encrypted, access is restricted to authorized personnel, and that data is stored securely. Additionally, organizations should have clear protocols in place for responding to data breaches, including notifying affected individuals and taking steps to mitigate the impact of the breach (Gallaher et al. 2023).

Informed Consent

Informed consent is another critical issue in the use of biometric technology in humanitarian settings. Individuals must fully understand what biometric data is being collected, how it will be used, and what risks are involved before they consent to its collection. However, obtaining informed consent can be challenging in humanitarian contexts, particularly when dealing with vulnerable populations who may not fully understand the implications of biometric data collection.

To address this issue, humanitarian organizations should prioritize transparency and communication, ensuring that individuals are fully informed about the use of their biometric data. This may involve providing information in multiple languages, using visual aids to explain complex concepts, and offering individuals the option to opt out of biometric data collection if they are uncomfortable with it (Fairhurst 2018).

Risk of Exclusion

There is also a risk that the use of biometric technology could lead to the exclusion of certain individuals from humanitarian assistance. For example, individuals with disabilities or injuries that affect their biometric identifiers may be unable to register in biometric systems, making it difficult for them to access aid. Additionally, technical issues with biometric systems, such as poor internet connectivity or equipment malfunctions, could prevent individuals from being properly identified and receiving assistance.

Humanitarian organizations must take steps to ensure that biometric systems are inclusive and accessible to all individuals, regardless of their physical or technical limitations. This may involve developing alternative identification methods for individuals who are unable to use biometric systems, such as manual registration or the use of proxy identifiers (Smith and Miller 2021).

Ethical Dilemmas in the Use of Biometrics

The use of biometrics in humanitarian operations also raises ethical questions. For instance, is it ethical to require refugees or other vulnerable populations to submit biometric data as a condition for receiving aid? What happens if individuals refuse to provide their biometric information due to cultural or religious beliefs?

These dilemmas are compounded by the fact that biometric technology is often implemented in crisis situations, where individuals may not have the capacity to fully understand or consent to the use of their data.

Legal and Regulatory Frameworks

The legal and regulatory frameworks governing the use of biometric technology in humanitarian contexts are still evolving. There is a need for clear guidelines and standards to ensure that biometric data is collected, stored, and used in a manner that respects the rights and dignity of individuals. International organizations, such as the United Nations, have begun to develop frameworks for the use of biometrics in humanitarian settings, but there is still much work to be done to ensure that these technologies are used responsibly (United Nations 2011).

Case Studies of Ethical Failures and Lessons Learned

Several incidents have highlighted the ethical challenges of using biometrics in humanitarian contexts:

Rohingya Refugee Crisis

One of the most prominent ethical controversies involving biometrics in recent years occurred during the Rohingya refugee crisis in Myanmar and Bangladesh. The United Nations High Commissioner for Refugees (UNHCR) implemented a biometric registration system in the Rohingya camps in Bangladesh, using fingerprints and iris scans to manage the distribution of aid and services. While this system significantly improved the efficiency of aid distribution, it also raised serious ethical concerns when reports surfaced that the biometric data collected was shared with the Myanmar government, potentially exposing refugees to further persecution upon their return (Ganguly 2021). This case underscores the need for strict data protection protocols and transparency in how biometric data is shared and used. The lesson learned here is that biometric data, especially in humanitarian contexts, must be safeguarded against misuse, with clear boundaries on how and with whom the data can be shared. The potential consequences of breaching these safeguards can be severe, including exposing vulnerable populations to harm from which they sought refuge.

Ebola Outbreak in West Africa

During the Ebola outbreak in West Africa (2014-2016), biometric technology was employed to monitor individuals who had been exposed to the virus. The use of fingerprint and facial recognition technology helped track the movement of individuals in and out of quarantine zones, ensuring compliance with public health measures. However, the implementation of these biometric systems raised concerns about privacy and stigmatization, as those identified through biometrics could potentially be ostracized or discriminated against by their communities (Yu and Carroll 2023). The ethical dilemma in this case lies in balancing public health needs with individual rights. The lesson learned is that while biometric systems can play a critical role in managing public health crises, they must be implemented with sensitivity to privacy concerns. There should also be measures in place to prevent the stigmatization of individuals identified through such systems, including robust communication strategies that educate communities on the purpose and benefits of biometric monitoring.

Jordan's Zaatari Camp

Jordan's Zaatari camp is one of the largest refugee camps in the world, housing over 76,000 Syrian refugees. The UNHCR has implemented an iris recognition system in the camp to manage the distribution of aid and services is often cited as a model of success. However, this case also presented ethical challenges related to consent and the potential exclusion of individuals who may be unable or unwilling to participate in the biometric registration process due to cultural or religious beliefs (Altare et al. 2022). The lesson from Zaatari is the importance of informed consent and inclusivity. Humanitarian organizations must ensure that all individuals fully understand the implications of biometric registration and that alternatives are provided for those who cannot or choose not to participate. This approach helps prevent exclusion and respects the cultural and religious values of the population being served.

Kenya's Dadaab Camp

Dadaab, located in Kenya, is one of the oldest and largest refugee camps globally, with a population of over 200,000. The WFP has implemented a fingerprint recognition system in Dadaab to manage food distribution. While the system significantly reduced fraud and improved the efficiency of aid delivery, it also highlighted ethical concerns about the handling of biometric data in environments where data security is often compromised (World Food Programme 2024). This case underscores the critical need for robust data security measures in humanitarian operations. The lesson learned is that in environments where infrastructure is often lacking, the risk of data breaches is higher, and therefore, organizations must prioritize data protection. This includes encrypting data, limiting access to authorized personnel, and having clear protocols for responding to security breaches.

The India Aadhaar System and Its Implications for Refugees

Although not exclusively a humanitarian operation, the Indian Aadhaar system's experience with refugees presents valuable lessons for the humanitarian sector. Aadhaar, a biometric identity system, has been criticized for its potential to exclude marginalized populations who lack the necessary documentation or who are unable to register due to physical limitations. The system's requirement for mandatory biometric registration has led to the exclusion of some of India's most vulnerable groups, including refugees and the elderly, from accessing essential services (Brinham and Johar 2024). The lesson from Aadhaar is that mandatory biometric registration can exacerbate social exclusion rather than alleviate it. Humanitarian organizations must therefore design biometric systems that are flexible and inclusive, ensuring that no individual is left without access to essential services due to their inability to comply with biometric requirements.

12.4 Future Prospects of Biometric Technology in Humanitarian Work

As we advance into an era marked by rapid technological innovation and growing global challenges, it is essential to explore the future prospects of biometric technology in humanitarian work. The future trajectory of biometric technology in humanitarian work is likely to be shaped by advancements in artificial intelligence (AI) and machine learning. AI-driven biometric systems can enhance the accuracy and efficiency of identification processes, providing real-time solutions to complex problems in crisis situations.

Potential Applications in Humanitarian Work

Refugee and Migrant Registration

Biometric systems offer significant advantages in managing refugee and migrant populations. By providing a unique and verifiable identity to individuals, these technologies can facilitate more effective distribution of aid, reduce fraud, and ensure that resources are allocated to those who need them most. Future advancements may allow for even more rapid and accurate processing, enhancing the overall efficiency of humanitarian operations.

Disaster Response and Recovery

In disaster-stricken areas, biometric technology can assist in identifying survivors, reuniting families, and ensuring the proper allocation of emergency resources. The integration of biometric data with geographical information systems (GIS) could further optimize the coordination of relief efforts, enabling targeted assistance based on the specific needs and locations of affected populations.

Healthcare Delivery

Biometric systems can also play a crucial role in healthcare delivery during crises. For instance, they can help track vaccination records, manage patient information in temporary clinics, and ensure that medical supplies are distributed based on accurate and up-to-date data. As biometric technology continues to improve, it may become an integral part of health information management systems in humanitarian settings.

Future Directions in Humanitarian Work

The future of biometric technology in humanitarian work holds exciting possibilities. Key areas for future development include:

Integration with Blockchain Technology

Combining biometric systems with blockchain technology could enhance data security and transparency, providing a tamper-proof record of identity and aid distribution.

Enhanced AI Algorithms

Continued advancements in AI could lead to more sophisticated biometric systems capable of operating in challenging environments and adapting to diverse populations.

Portable and Affordable Solutions

Developing portable and cost-effective biometric solutions will be crucial for ensuring that technology can be deployed quickly and widely in emergency situations.

Future Directions in Biometric Technologies

In the realm of humanitarian work, several key biometric technologies are emerging or advancing, each with its unique applications and potential benefits. Below are the major biometric technologies that could be utilized:

Fingerprint Recognition

- Overview: This technology analyzes the unique patterns of ridges and valleys on an individual's fingertips.
- Applications: Commonly used for identity verification in refugee registration and secure access to aid. It's robust and relatively inexpensive, making it suitable for large-scale deployments.

Facial Recognition

- Overview: Facial recognition systems analyze the unique features of an individual's face, such as the distance between eyes, nose, and mouth.
- Applications: Useful for quickly identifying individuals in crowds or verifying identities at checkpoints. It can be employed for both registration and monitoring, but requires good lighting and high-quality images for accuracy.

Iris Recognition

- Overview: This technology captures and analyzes the unique patterns in the iris of the eye.
- Applications: Highly accurate and stable over time, making it suitable for verifying identities in sensitive or high-security contexts. Iris recognition can be used for managing access to medical services or distributing aid.

Voice Recognition

- Overview: Voice recognition systems analyze the unique characteristics of an individual's voice, including pitch, tone, and speech patterns.
- Applications: Can be used for remote verification of identities and providing access to telephonic or virtual services. It's less intrusive but can be affected by background noise or health conditions affecting the voice.

Vein Recognition

- Overview: This technology analyzes the pattern of veins in an individual's hand or finger using infrared light.
- Applications: Offers a high level of security and is difficult to forge. It's useful for identity verification in environments where high accuracy is required, such as in financial transactions or secure facilities.

Behavioral Biometrics

- Overview: This includes technologies that analyze behavioral patterns such as typing rhythm, mouse movements, or gait.
- Applications: Can be used for ongoing authentication and fraud prevention, particularly in digital interactions. It's useful for monitoring and verifying user identity in online platforms.

DNA Recognition

- Overview: Analyzes the unique genetic material of an individual.
- Applications: Highly accurate and can be used for identifying individuals in extreme cases, such as disaster victim identification. However, it is more resource-intensive and time-consuming compared to other biometric methods.

Hand Geometry

- Overview: Measures and analyzes the shape and size of an individual's hand and fingers.
- Applications: Useful for access control and identity verification in situations where users can be physically present. It's less precise than fingerprint or iris recognition but can be effective in specific contexts.

4.3.9.Gait Recognition

- Overview: Analyzes the way an individual walks.
- Applications: Can be used for identifying individuals at a distance and in motion, making it useful in security and surveillance contexts. It's less commonly used but holds potential for certain humanitarian applications.

Thermal Imaging

- Overview: Uses heat signatures to identify individuals based on their unique thermal patterns.
- Applications: Useful in low-visibility conditions, such as during night-time or in smoke-filled environments. It can assist in locating individuals or verifying identities in challenging conditions.

Integration and Future Trends in Humanitarian Work

The integration of these biometric technologies with other emerging technologies such as artificial intelligence (AI), blockchain, and the Internet of Things (IoT) is likely to enhance their effectiveness in humanitarian work. For instance:

AI and Machine Learning

AI can improve the accuracy and efficiency of biometric systems by enabling more sophisticated pattern recognition and reducing error rates.

Blockchain

Combining biometric data with blockchain technology can provide secure, tamper-proof records of identity and aid distribution.

Portable Solutions

Advances in miniaturization and mobile technology are making biometric systems more portable and accessible, which is crucial for deployment in remote or disaster-stricken areas.

Overall, the choice of biometric technology will depend on the specific needs and constraints of the humanitarian context, including considerations of accuracy, cost, and the environment in which the technology will be used.

12.5 Conclusion

The integration of biometric technology into humanitarian work presents both significant opportunities and complex challenges. As the technology continues to evolve, it has the potential to transform the way humanitarian organizations manage crises, deliver aid, and protect vulnerable populations. By addressing the ethical, privacy, and technological challenges associated with biometric systems, we can harness their full potential to enhance the effectiveness and efficiency of humanitarian efforts. Future research and innovation in this field will be critical in shaping a future where biometric technology contributes positively and meaningfully to global humanitarian endeavors.

References

Altare C, Kostandova N, OKeeffe J, Hayek H, Fawad M, Musa Khalifa A, Spiegel PB (2022) COVID-19 epidemiology and changes in health service utilization in Azraq and Zaatari refugee camps in Jordan: a retrospective cohort study. PLoS Med 19(5). https://doi.org/10.1371/journal.pmed.1003993

Brinham N, Johar A (2024 May) Refugee experiences of identity documents and digitisation in India and Myanmar—forced migration review. https://www.fmreview.org/digital-disruption/brinham-johar/

Carrillo E, Sequera M (2020) Personal data in the social security institute. Law, State Telecommun Rev 12(2). https://doi.org/10.26512/lstr.v12i2.34689

Fairhurst M (2018) Biometrics: a very short introduction. https://doi.org/10.1093/actrade/9780198809104.001.0001

Francese R, Frasca M, Risi M (2021) Are IoBT services accessible to everyone? Pattern Recognit Lett 147. https://doi.org/10.1016/j.patrec.2021.04.007

Gallaher MP, Link AN, Rowe BR (2023) Vulnerabilities and cyber security technologies. Cyber Secur 33–59. https://doi.org/10.4337/9781781008140.00010

Ganguly M (2021 June 7) Bangladesh's unplanned relocation of Rohingya refugees to Bhasan char Island is risky | human rights watch. https://www.hrw.org/news/2021/06/07/bangladeshs-unplanned-relocation-rohingya-refugees-bhasan-char-island-risky

Gavrilova ML (2008) Exploring fingerprint matching through a topology-based perspective. Int J Biom 1(1). https://doi.org/10.1504/IJBM.2008.018661

Gelb A, Clark J (2013) Identification for development: the biometrics revolution. SSRN Electron J. https://doi.org/10.2139/ssrn.2226594

Mackey, T. C., & Cole, S. A. (2001). Suspect identities: a history of fingerprinting and criminal identification. Am J Leg Hist 45(3). https://doi.org/10.2307/3185338

Smith M, Miller S (2021) Biometric identification, law and ethics. Springerbriefs in ethics

Tran QN, Turnbull BP, Hu J (2021) Biometrics and privacy-preservation: how do they evolve? IEEE Open J Comput Soc 2. https://doi.org/10.1109/ojcs.2021.3068385

UNHC (2018) UNHCR—UNHCR global trends 2018. https://www.unhcr.org/statistics/unhcrstats/5d08d7ee7/unhcr-global-trends-2018.html

UNHCR (2024a) Bangladesh | UNHCR. https://www.unhcr.org/countries/bangladesh

UNHCR (2024b) Refugee Camps—UNHCR Jordan. https://www.unhcr.org/jo/refugee-camps

United Nations (2011) Guiding principles for business and human rights: implementing the United Nations "protect, respect and remedy" framework | UN global compact. https://unglobalcompact.org/library/2

Upadhyay J, Paranjpe R, Purohit H, Joshi R (2020) Biometric identification using gait analysis by deep learning. 2020 IEEE Pune section international conference, PuneCon 2020. https://doi.org/10.1109/PuneCon50868.2020.9362402

World Food Programme (2024) UN World Food Programme (WFP). https://www.wfp.org/

Yu S, Carroll F (2023) Securing privacy during a world health emergency: exploring how to create a balance between the need to save the world and people's right to privacy. Data Prot Post-Pandemic Soc. https://doi.org/10.1007/978-3-031-34006-2_5

Nurul Ain Safwah is an academician and program coordinator, actively involved in research field of maritime operation, transportation, and logistics. She is currently attached to Akademi Laut Malaysia. Her expertise contributes significantly to the development of maritime education in Malaysia. Safwah's research focuses on improving efficiency and sustainability in maritime operations, maritime humanitarian effort and addressing crucial industry challenges.

Ramesh Babu is an experienced lecturer and currently leading the Faculty of Business & Management with over 25 years of experience in Information Technology and Information System, together with business programs. Ramesh Babu has a strong history of success across multiple institutions, demonstrating deep expertise and acommitment to inspiring and effective teaching. His approach combines extensive knowledge with a passion for education, ensuring meaningful learning experiences for students.

13 Leveraging Explainable AI for Optimizing Humanitarian Aid Distribution: Lessons from Flight Delay Prediction in Air Traffic Management

Daniel D. Rumani and Darmeli Nasution

Abstract

This research examines the application of machine learning and Explainable AI (XAI) methods to predict the distribution of flight delays in the context of air traffic management. With the increasing complexity and volume of data in the aviation industry, accurate delay predictions are key to improving operational efficiency and safety. This research explores various machine learning algorithms, including Support Vector Machine (SVM) and Random Forest, to model and predict flight delays based on historical data and influencing factors, such as weather conditions, air traffic, crew availability, and delay times. The results of this research are machine learning exploration in the form of support vector machine and random forest algorithms in predicting air traffic delays resulting in a model with good accuracy between the support vector machine and random forest algorithms. The support vector machine algorithm produces an accuracy of 0.9, while the random forest algorithm produces an accuracy of 0.8. and the application of Explainable AI (XAI) produces variables that can explain delay factors. By using XAI, you can identify which variables have the greatest influence on flight delays. The result is that when implementing XAI, weather variables are more influential, so they are able to predict more accurately when bad weather occurs.

Keywords

Prediction · XAI · SVM · Random forest · Air traffic

13.1 Introduction

Air traffic management is the most influential part in ensuring the safety and regularity of flight operations and also influences efficiency. In the aviation industry, which is the most developed industry, there will be challenges faced, especially in terms of flight delays (Degas et al. 2022; Pineda-Jaramillo et al. 2024). Delays will be influenced by many factors including weather conditions, air traffic density, machine problems, and technical errors made by humans. On the other hand, delays in flights will affect customer satisfaction and financial losses experienced by airlines (Dalmau 2024; Kravaris et al. 2023). So the problem in this research is that there are

D. D. Rumani (✉)
Akademi Penerbang Indonesia,
Banyuwangi, Indonesia
e-mail: danieldewantoro@icpa-banyuwangi.ac.id

D. Nasution
Universitas Pembangunan Pancabudi,
Medan, Indonesia
e-mail: darmelinasution@gmail.com

N. A. A. Rahman et al. (eds.), *Supply Chain in Humanitarian Operations*, Sustainable Development Goals Series,
https://doi.org/10.1007/978-981-96-4426-1_13

many delays that occur which require applying prediction techniques aimed at flight delays in air traffic management. This prediction will concern the distribution of delays so that it can be used to take mitigation steps (Kistan et al. 2018).

AI and Machine Learning in Humanitarian Supply Chain

AI and Machine Learning have the potential to revolutionize humanitarian efforts. For instance, AI can analyze large datasets to identify patterns and trends, helping humanitarian workers understand the needs of communities affected by disasters. Furthermore, AI can be used to develop predictive models that enable humanitarian workers to anticipate future needs and plan effective responses. Additionally, machine learning can be used to develop algorithms that optimize the use of resources. For example, machine learning can create algorithms to determine the most efficient routes for delivering aid or to allocate resources fairly. Overall, AI and machine learning have the potential to make humanitarian work more effective and efficient. By utilizing these technologies, humanitarian workers can better understand the needs of affected communities, predict future needs, and plan more effective responses (Mohammad et al. 2022).

In line with technological advances, there will be a branch of the discipline of artificial intelligence, namely machine learning, which is a technique for making predictions so that it can be used to predict flight delays. Machine learning has the ability to process and analyze big data quickly and efficiently, enabling the identification of complex patterns that are difficult to see by traditional statistical methods. In the context of air traffic management, machine learning can be used to predict factors that influence flight delays based on historical data, such as flight schedules, weather, and airport capacity (Ai et al. 2019; Pinto Neto et al. 2023). Thus, the resulting predictions can help decision-makers to manage resources more effectively and minimize delays. In this context, predicting flight delays will involve an effective algorithm in making predictions. This research will use support vector machine, random forest, and decision tree algorithms. In this case, each algorithm has its own advantages and disadvantages. In this research, there are challenges in implementing machine learning algorithms, namely the quality of the data used, such as weather data, air traffic data, and airport condition data which must be well structured (Truong 2021; Hernandez et al. 2021).

A number of previous studies have explored the use of machine learning in the context of flight delays. For example, research conducted by Zhang et al. (2020) used the random forest method to predict flight delays at large airports in the United States. The results of this research show that flight delay predictions can be made with fairly high accuracy, especially when weather and air traffic data are integrated into the model. Another study by Alharbi et al. (2023) examined the influence of weather factors on flight delays using a logistic regression model and found that bad weather was one of the main factors causing delays. In implementing machine learning in carrying out the task of predicting flight delays, there are challenges in how the model performs its task. The algorithms used are difficult for humans to understand or interpret. So techniques are needed to understand the working model to make decisions in predicting delays with the aim of airport management, especially airlines, knowing what factors are most influential in influencing delays. Explainable AI (XAI) technology will be used to provide explanations to machine learning models so that users can understand the decisions produced by the system (Gui et al. 2020; Yazdi et al. 2020). The XAI approach helps visualize key factors that contribute to delays, such as weather conditions, air traffic, and aircraft reliability. So this research will explore three machine learning algorithms that will be applied in predicting flight delays and will then also apply XAI techniques to see what factors influence delays (Yu et al. 2019; Shao et al. 2022).

13.2 Literature Review

A. Machine Learning

Machine learning is a part of artificial intelligence (AI) that focuses on developing algorithms and models that allow computers to learn from data and make predictions or decisions without being explicitly programmed. With machine learning, computers can recognize patterns and trends in datasets and use this information to solve problems or make decisions in future. This learning process usually involves training a model using some historical data, and then using the model to make predictions based on new data (Lubis et al. 2023). Machine learning is used in applications such as voice recognition, image recognition, data analysis, and product recommendations, as well as in the fields of health, finance, and transportation (Arfah et al. 2024). Machine learning is divided into several main categories, namely supervised learning, unsupervised learning, and reinforcement learning. In supervised learning, the model is trained using a labeled dataset, where the input and output are known. A common example of supervised learning is house price prediction based on features such as land area, number of rooms, and location, where historical house price data is used as a reference. Algorithms such as linear regression, Support Vector Machine (SVM), and Random Forest fall into this category. In contrast, unsupervised learning does not require labeled data. Models in unsupervised learning try to find hidden structures in data, such as clustering and dimensionality reduction. Algorithms such as K-Means and Principal Component Analysis (PCA) are often used to analyze hidden patterns in unstructured datasets.

B. Air Traffic Management

Air traffic management is a set of procedures, technologies, and services used to ensure that the movement of aircraft in the air and on the ground is safe, efficient, and orderly. This management includes monitoring, controlling, and regulating aircraft traffic in congested air spaces or around airports (Xie et al. 2021). The main goal of air traffic management is to prevent collisions between aircraft, manage distances between aircraft, and optimize flight paths to fit schedules and reduce delays. This function is usually performed by Air Traffic Control (ATC), which is responsible for providing instructions to the pilot regarding altitude, direction, and speed during flight as well as during takeoff or landing. Air traffic management also includes monitoring weather, heavy air traffic, and other operational factors to ensure smooth flight operations (Liu et al. 2019).

C. XAI (Explainable Artificial Intelligence)

Explainable Artificial Intelligence (XAI) is a branch of artificial intelligence (AI) that focuses on making AI models more transparent, understandable, and explainable to humans. The main goal of XAI is to enable users or developers to understand how and why an AI model makes certain decisions or predictions. XAI is critical in increasing trust in AI systems, especially in critical applications such as healthcare, finance, and law, where understanding the decision-making process is critical (Mehta and Passi 2022). With XAI, users can evaluate whether the decisions produced by the model are fair, accurate, and free from bias. XAI provides tools and techniques to reveal the internal processes of AI models that are often complex, especially in models such as deep learning or machine learning which are often considered "black boxes." This approach may include visualizations, rules that can be followed, or statistics-based explanations to provide insight into the predictions made by the model (Temenos et al. 2022).

D. Prediction

Prediction is the process of estimating or predicting an event or outcome that will occur in future based on currently available data or information. In the context of technology and data, prediction often involves the use of algorithms, statistical models, or machine learning to analyze patterns in historical data to project future trends or events (Rahmatika et al. 2024). Forecasting in air traffic is the process of estimating various factors that may affect flight operations, such as flight delays, air congestion, weather conditions, or future availability of air space. This prediction is carried out by analyzing historical data and real-time data using algorithms and machine learning models to project potential disturbances or events that may occur in aircraft movements. With accurate predictions, airline operators, air traffic controllers, and airports can make better decisions to avoid delays, optimize flight routes, and improve operational efficiency and safety (Jha et al. 2020).

13.3 Methodology

This section will explain the methodology that will be used to apply machine learning algorithms and techniques from Explainable AI (XAI) in predicting the distribution of flight delays in air traffic management. In this context, flight data will be collected which includes information on flight schedules, weather, and airport capacity. After that, data preprocessing is carried out with the aim of making the data more structured so that it is easily processed by machine learning algorithms. After that, support vector machine and random forest algorithms will be applied with the aim of comparing machine learning algorithms that are effective in predicting flight delays, then integrating techniques from Explainable AI (XAI) to ensure that the prediction results can be understood and interpreted clearly by airport managers and airlines. XAI methods like LIME (Local Interpretable Model-agnostic Explanations) will be used to explain how certain factors influence prediction results and provide deeper insight into the model's decision-making. Then the evaluation stage will use evaluation metrics such as accuracy, precision, recall, and F1 score. In addition, to assess the model's ability to correctly predict delay distribution, the AUC-ROC (Area Under the Receiver Operating Characteristic Curve) curve will be used which provides an overview of the trade-off between true positive rate and false positive rate. This evaluation is important to ensure that the model is not only accurate but also provides useful interpretations for stakeholders in air traffic management.

Table 13.1 Dataset

Delay_in_Minutes	Weather_Conditions	Air_Traffic	Crew_Availability
−22	Rain	High	Available
10	Rain	Low	Available
−8	Fog	Low	Available
−8	Storm	High	Available
36	Fog	Low	Delayed
88	Fog	Low	Delayed
12	Clear	Low	Available
73	Fog	Low	Available

A. Dataset

At this stage we will use a dataset with a total of 2000 data. At this stage data processing will be carried out to eliminate data that has no value. In this data there will be columns for delay time, weather conditions, traffic density. The following data is used in Table 13.1.

B. Research Architecture

In this section there will be stages in applying machine learning algorithms and XAI to predict air traffic delays. At this stage, we will explain the use of support vector machine and random forest algorithms, then XAI will be applied to explain how important the variables are that influence delays. The following is the research architecture in Fig. 13.1.

In Fig. 13.1 there will be a research architecture for predicting delays which will be explained as follows:

1. At this stage we will use 4 parameters that will be used to predict delays;
2. Perform data cleaning which aims to eliminate empty data and unstructured data;
3. Build a prediction model by exploring machine learning algorithms such as support

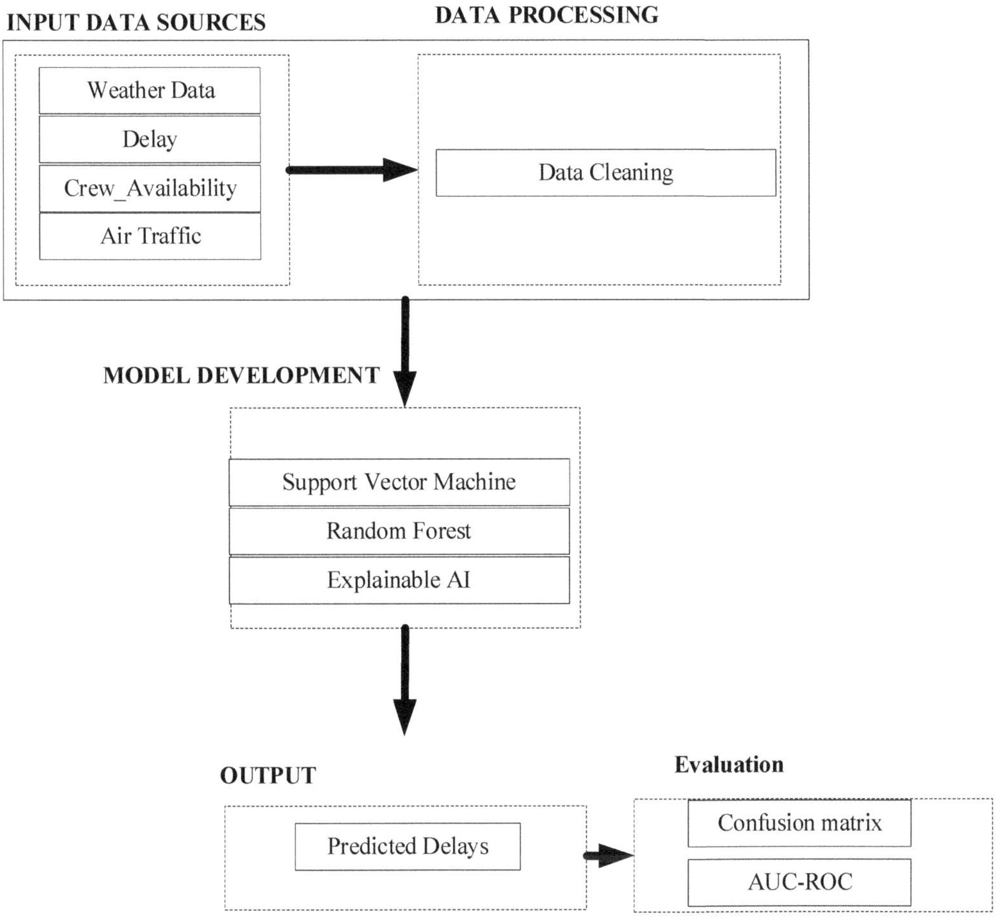

Fig. 13.1 Research architecture

vector machines and random forests which will then be applied by XAI techniques to see what parameters have the most influence on air traffic delays;
4. Conduct model evaluation using confusion matrix and AUC-ROC evaluation techniques.

13.4 Discussion

This section is a discussion section related to predicting air traffic delays by exploring machine learning in the form of support vector machine algorithms and random forest algorithms. Each algorithm has advantages and disadvantages in time complexity and accuracy. In this case the model will carry out a training process and a testing process which will be evaluated using the AUC-ROC evaluation technique. The model will apply XAI which is part of the explanation of the results that allows humans to understand them. In this context, we will explain what parameters are very influential in predicting delays. By knowing these parameters, they will be used for decision-making in air traffic management. In this case the model will apply explanatory data analysis to describe the data so that it can be understood. The following is a discussion starting from explanatory data analysis to model evaluation.

A. Data

At this stage, explanatory data analysis will display graphs on the data that will describe the data directly so that you can see what parameters are used and how much data. The following is a graph of the data with the delay parameters in Fig. 13.2.

Figure 13.2 will explain the delay parameters with the amounts depicted via graph. Apart from that, there will be graphs in the form of plots for all parameters such as air traffic, weather conditions, crew availability at the airport which will be displayed in Fig. 13.3.

Figure 13.3 will explain the data distribution based on each variable, then there will be a correlation heatmap which is a graphical representation of the correlation matrix between the variables in the dataset. Typically, these heatmaps are used to show the relationships between various variables in a way that is easy to interpret visually. The following is the correlation heatmap in Fig. 13.4.

B. Model Prediction

In this section, the results of the air traffic delay prediction model will be displayed. In this section the model is successfully implemented by exploring machine learning algorithms such as support vector machines and random forest algorithms. Each algorithm can work well resulting in different accuracy. In the context of machine learning algorithm exploration, training will be carried out and then model testing and evaluation will be carried out to see the performance of the air traffic delay prediction model. In this section, there will be tests for predicting delays in each algorithm as follows:

(a) Support Vector Machine

In this section, testing is carried out using a support vector machine algorithm with training data divided into 80% of the total data and testing 20% of the total data. The following are the test results in Table 13.2. Figure 13.6 explains about the confusion matrix random forest.

Based on Table 13.2, there will be 400 data that will be tested on the support vector machine algorithm which produces an accuracy of 0.9 in predicting air traffic delays with the variables used. Then there will be an evaluation matrix to see the performance of the model in Fig. 13.5.

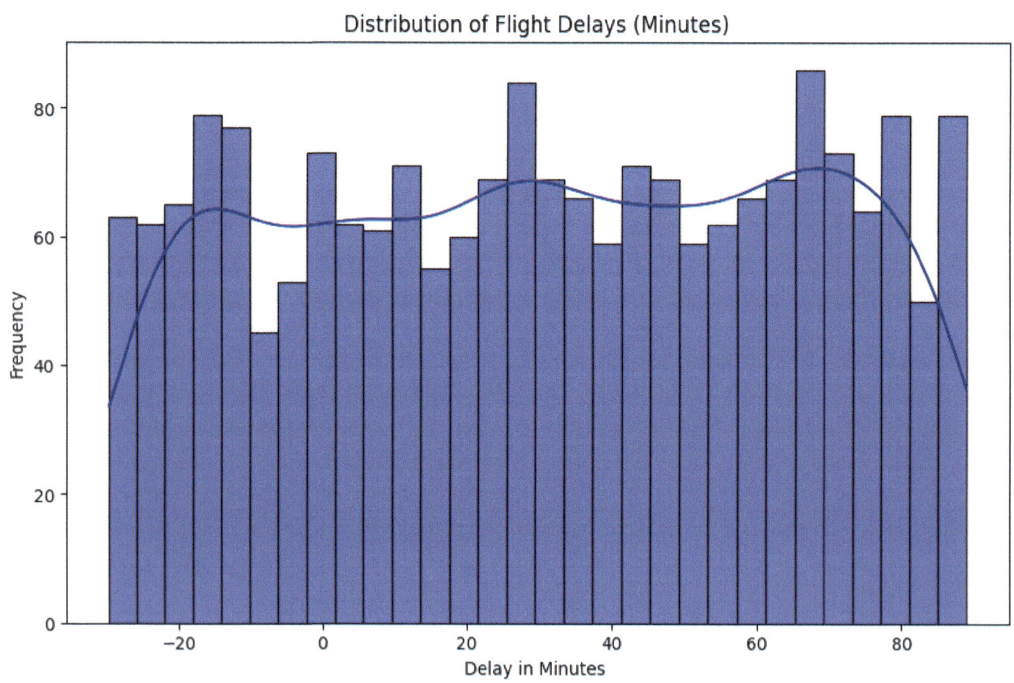

Fig. 13.2 Parameters flight delay Distribution

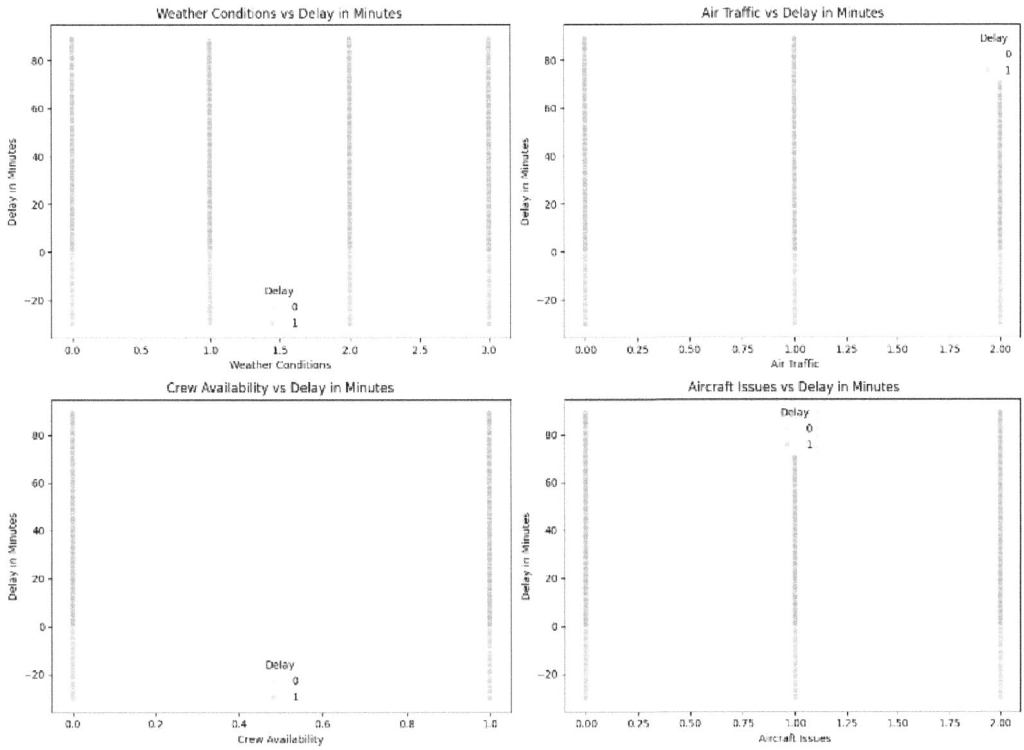

Fig. 13.3 Distribution of parameters

(1) Random Forest

In this section, testing is carried out using a random forest algorithm with training data divided into 80% of the total data and testing 20% of the total data. The following are the test results in Table 13.3.

Based on Table 13.3, there will be 400 data that will be tested on the random forecast algorithm which produces an accuracy of 0.84 in predicting air traffic delays with the variables used. Then there will be an evaluation matrix to see the performance of the model in Fig. 13.5.

C. Comparison of SVM and Random Forest Algorithms

In predicting air traffic delays, the Support Vector Machine (SVM) and Random Forest algorithms each have significant advantages and disadvantages. SVM, with its ability to find the optimal hyperplane and separate classes with maximum margin while Random Forest is also more resistant to variability and noise in the data, making it suitable for situations where many factors interact with each other, such as in the analysis of air traffic delays which are influenced by variables such as weather conditions, air traffic, crew availability, and delay times. The following Image shown in Fig. 13.7 is a graph comparing the accuracy between the SVM and random forest algorithms.

D. Evaluation Curves ROC (Receiver Operating Characteristic)

In this section, an evaluation model will be carried out using the AUC-ROC technique to see the performance of the model in predicting delays. In this case, AUC-ROC produces a value of 0.96, which concludes that the model can work better in predicting delays. The model evaluation will calculate the TPR value and FPR value on the AUC-ROC graph. The

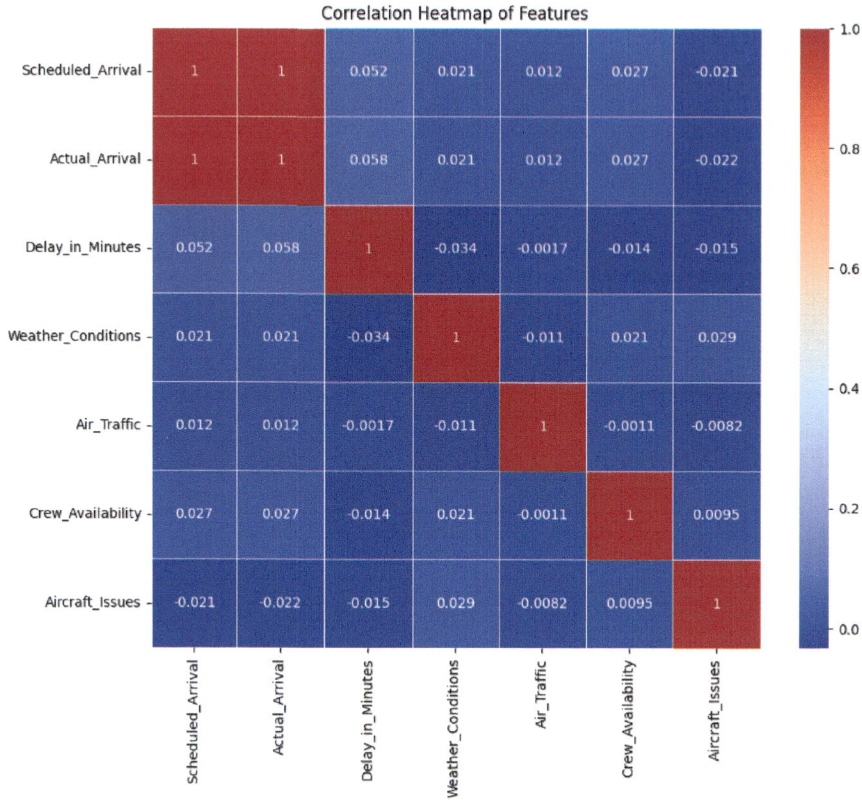

Fig. 13.4 Correlation heatmap

Table 13.2 SVM testing algorithm

No	Label	Prediction
1	Delay	Delay
2	Delay	Delay
3	Non delay	Non delay
4	Delay	Delay
5	Delay	Delay
399	Delay	Delay
400	Non delay	Non delay

following are the results of the model evaluation in Fig. 13.8.

E. Evaluation Curves ROC (Receiver Operating Characteristic)

Explainable Artificial Intelligence (XAI) in the context of flight delay prediction can provide transparent insight into the variables that influence prediction results, such as weather conditions, air traffic, crew availability, and delay times. Bad weather is often the main cause of delays, so machine learning models can identify extreme weather patterns that increase the likelihood of delays. Heavy air traffic, especially at large airports or popular routes, also affects schedule accuracy because air space is limited to manage high flight volumes. On the other hand, crew availability plays an important role in ensuring smooth flight operations. Models that are able to explain the relationship between low crew availability and increased risk of delays can help airlines optimize scheduling. The final factor, namely delay time, functions as a variable that can be predicted and explained in relation to other external conditions, allowing the model to provide clear reasons for delays. The XAI approach ensures that the decisions taken by the model

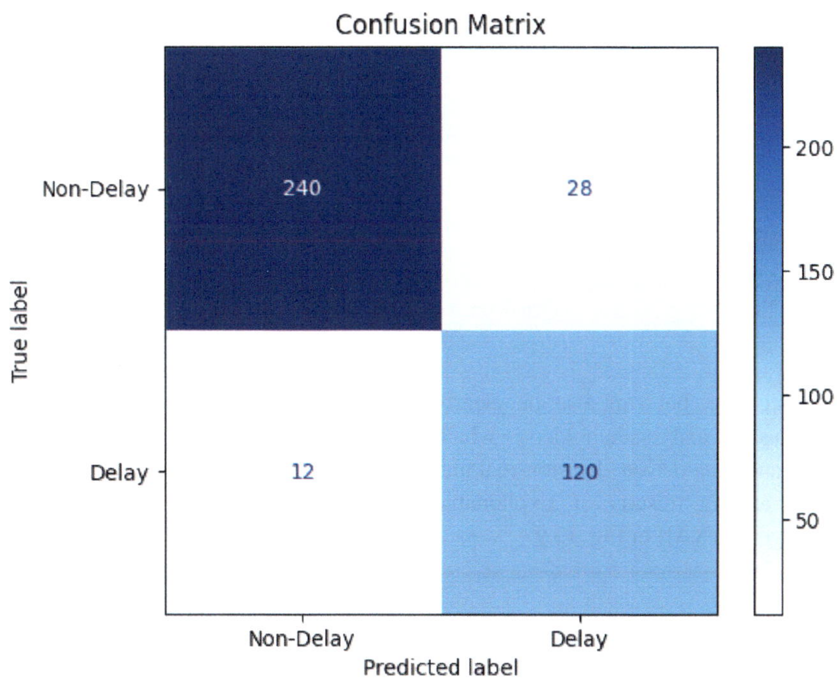

Fig. 13.5 Confusion matrix support vector machine

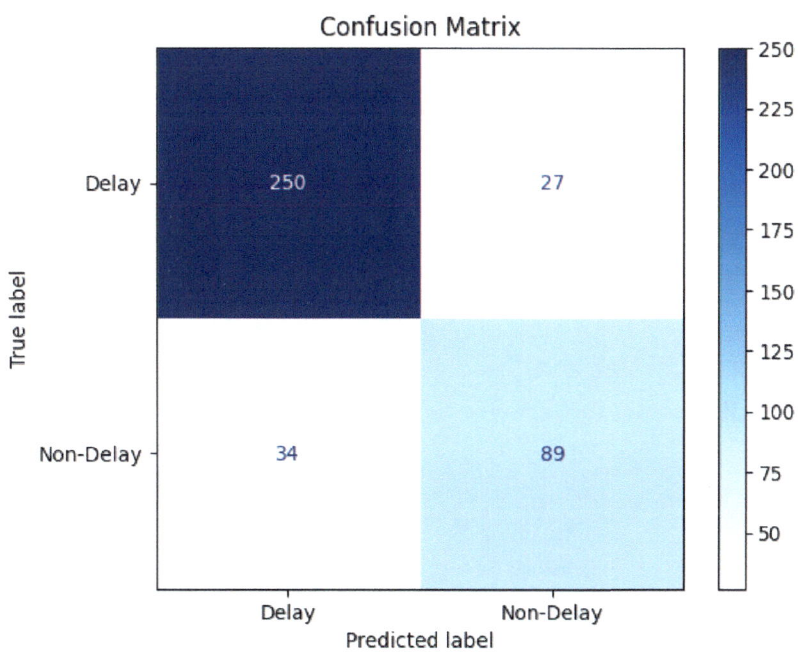

Fig. 13.6 Confusion matrix random forest

Table 13.3 Testing algorithm random forest

No	Label	Prediction
1	Non Delay	Delay
2	Delay	Delay
3	Non delay	delay
4	Non Delay	Non Delay
5	Delay	Delay
399	Delay	Delay
400	Non delay	Non delay

are not only accurate, but can also be clearly explained to users and stakeholders, which increases trust and data-driven decision-making. The following are the results of Explainable Artificial Intelligence (XAI) in Fig. 13.9.

13.5 Conclusion

Exploration of machine learning in the form of support vector machine and random forest algorithms in predicting air traffic delays produces a model with good accuracy between the support vector machine and random forest algorithms. The support vector machine algorithm produces an accuracy of 0.9, while the random forest algorithm produces an accuracy of 0.8. This related difference in accuracy indicates that the SVM algorithm is slightly superior in predicting air traffic delays compared to Random Forest. SVM is able to distinguish categories more precisely in a high-dimensional space, making it an ideal choice for data that has clear margins between classes. On the other hand, Random Forest offers advantages in terms of stability and ability to handle data with high variability, although its accuracy is slightly lower. In this context, the role of Explainable Artificial Intelligence (XAI) is very important to understand how the two algorithms produce predictions and the differences in accuracy. XAI enables deeper interpretation of the variables that contribute to predictions, such as weather conditions, air traffic, crew availability, and delay times. By using XAI, you can identify which variables have the greatest influence on flight delays. The result is that when implementing XAI, weather variables are more influential, so they are able to predict more accurately when bad weather occurs.

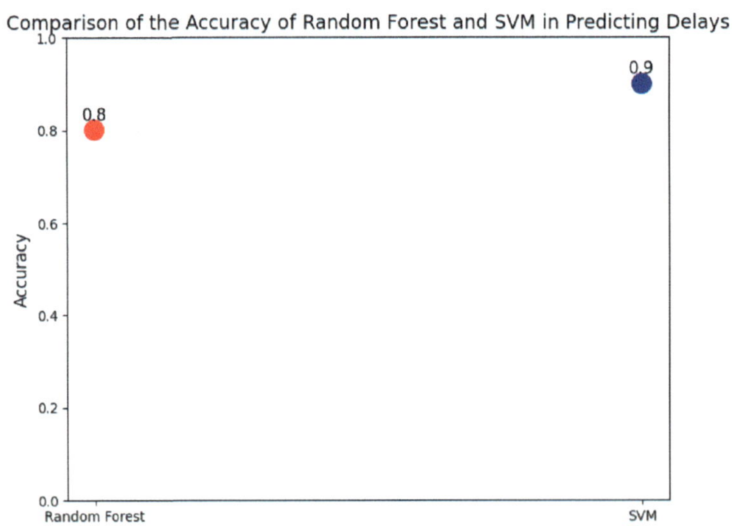

Fig. 13.7 Comparison plot of test results

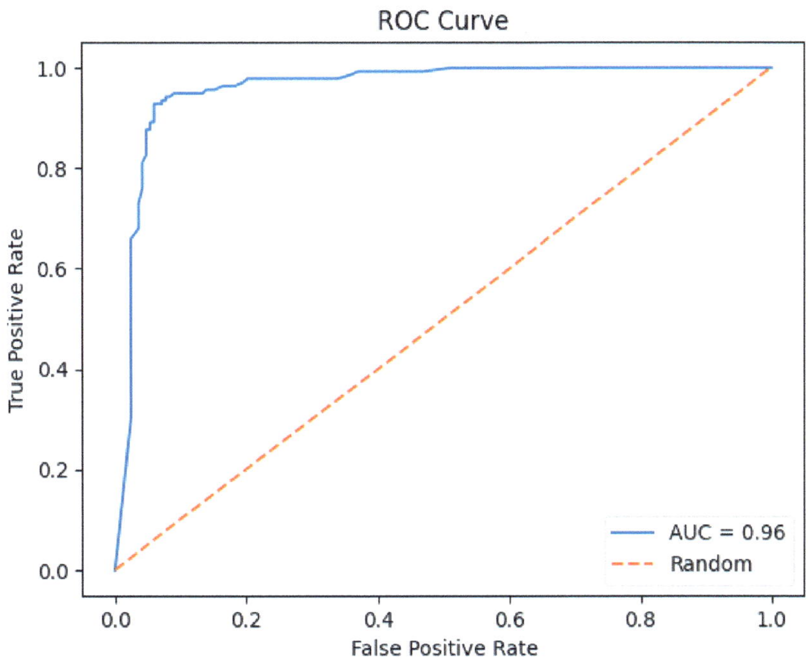

Fig. 13.8 Result AUC-ROC

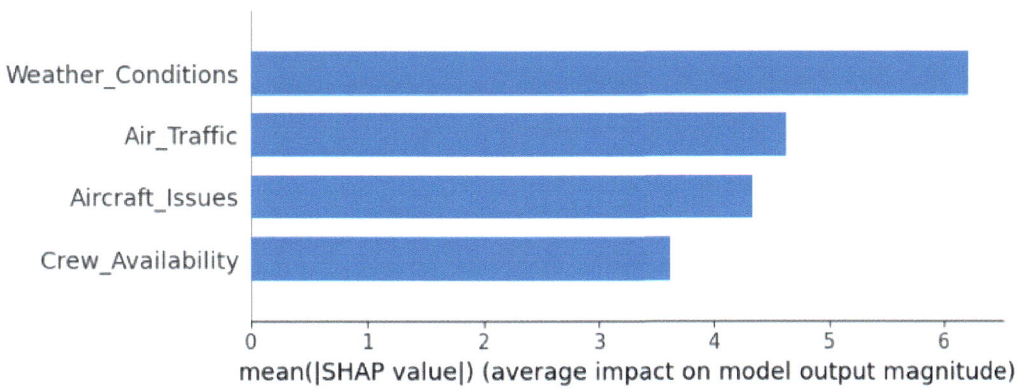

Fig. 13.9 Result Explainable Artificial Intelligence (XAI)

References

Ai Y, Pan W, Yang C, Wu D, Tang J (2019) A deep learning approach to predict the spatial and temporal distribution of flight delay in network. J Intell & Fuzzy Syst 37(5):6029–37

Alharbi A, Petrunin I, Panagiotakopoulos D (2023) Assuring safe and efficient operation of uav using explainable machine learning. Drones 7(5):327

Arfah M, Fachrizal F, Nugroho O (2024) Developing a model of association rules with machine learning in determining user habits on social media. 55–61. https://doi.org/10.15587/1729-4061.2024.305116

Dalmau R (2024) Probabilistic and explainable tree-based models for rotational reactionary flight delay prediction. CEAS Aeronaut J 1–17

Degas A, Islam MR, Hurter C, Barua S, Rahman H, Poudel M, Ruscio D, Ahmed MU, Begum S, Rahman

MA (2022) A survey on Artificial Intelligence (Ai) and explainable Ai in air traffic management: current trends and development with future research trajectory. Appl Sci 12(3):1295

Gui G, Zhou Z, Wang J, Liu F, Sun J (2020) Machine learning aided air traffic flow analysis based on aviation big data. IEEE Trans Veh Technol 69(5):4817–26

Hernandez CS, Ayo S, Panagiotakopoulos D (2021) An Explainable Artificial Intelligence (XAI) framework for improving trust in automated ATM tools. In: 2021 IEEE/AIAA 40th Digital Avionics Systems Conference (DASC). IEEE, pp 1–10

Jha G, Ahuja L, Rana A (202) Criminal behaviour analysis and segmentation using k-means clustering. In: ICRITO 2020—IEEE 8th international conference on reliability, infocom technologies and optimization (trends and future directions), pp 1356–60. https://doi.org/10.1109/ICRITO48877.2020.9197791

Kistan T, Gardi A, Sabatini R (2018) Machine learning and cognitive ergonomics in air traffic management: recent developments and considerations for certification. Aerospace 5(4):103

Kravaris T, Lentzos K, Santipantakis G, Vouros GA, Andrienko G, Andrienko N, Crook I, Garcia JMC, Martinez EI (2023) Explaining deep reinforcement learning decisions in complex multiagent settings: towards enabling automation in air traffic flow management. Appl Intell 53(4):4063–98

Liu Y, Liu Yi, Hansen M, Pozdnukhov A, Zhang D (2019) Using machine learning to analyze air traffic management actions: ground delay program case study. Transp Res Part E: Logist Transp Rev 131:80–95

Lubis AR, Safitri HR, Lubis M, Hamzah ML, Al-Khowarizmi AK, Nugroho O (2023) Enhancing text summarization with a T5 model and Bayesian optimization. Rev d'Intelligence Artif 37(5)

Mehta H, Passi K (2022) Social media hate speech detection using Explainable Artificial Intelligence (XAI). Algorithms 15(8). https://doi.org/10.3390/a15080291

Mohammad S, Bhowmick T, Siddique MSUZ, Khan MM, Bhattacharyya S (2022) Research and development of a artificial intelligence based smart medicine box. In: 2022 6th International Conference on Computing Methodologies and Communication (ICCMC). IEEE, pp. 407–12

Pineda-Jaramillo J, Munoz C, Mesa-Arango R, Gonzalez-Calderon C, Lange A (2024) Integrating multiple data sources for improved flight delay prediction using explainable machine learning. Res Transp Bus Manag 56:101161

Neto P, Carlos E, Baum DM, Rady J, de Almeida Jr, João Batista Camargo Jr, and Paulo Sergio Cugnasca. (2023) Deep learning in Air Traffic Management (ATM): a survey on applications, opportunities, and open challenges. Aerospace 10(4):358

Rahmatika A, Nugroho O, Anu TA (2024) Using relational learning in exploring the effectiveness of using hashtags in future topics and user relations in X. 2:62–68. https://doi.org/10.15587/1729-4061.2024.306726

Shao W, Prabowo A, Zhao S, Koniusz P, Salim FD (2022) Predicting flight delay with spatio-temporal trajectory convolutional network and airport situational awareness map. Neurocomputing 472:280–93

Temenos A, Tzortzis IN, Kaselimi M, Rallis I, Doulamis A, Doulamis N (2022) Novel insights in spatial epidemiology utilizing explainable AI (XAI) and remote sensing. Remote Sensing 14(13):3074

Truong D (2021) Using causal machine learning for predicting the risk of flight delays in air transportation. J Air Transp Manag 91:101993

Xie Y, Pongsakornsathien N, Gardi A, Sabatini R (2021) Explanation of machine-learning solutions in air-traffic management. Aerospace 8(8):224

Yazdi MF, Kamel SR, Chabok SJM, Kheirabadi M (2020) Flight delay prediction based on deep learning and levenberg-marquart algorithm. J Big Data 7:1–28

Yu B, Guo Z, Asian S, Wang H, Chen G (2019) Flight delay prediction for commercial air transport: a deep learning approach. Transp Res Part E: Logist Transp Rev 125:203–21

Zhang K, Jiang Y, Liu D, Song H (2020) Spatio-temporal data mining for aviation delay prediction. in 2020 IEEE 39th International Performance Computing and Communications Conference (IPCCC). IEEE, pp. 1–7

Daniel Dewantoro Rumani is an associate professor and also a director at Indonesian Pilot Academy Banyuwangi, commenced their career as a lecturer in 2001 as per Ministerial Decree No. 91 of 2001 concerning the appointment of Extraordinary Lecturers at Higher Education Institutions under the Ministry of Transportation. In 2012, they served as an Aircraft Operations Inspector, Sub-Directorate of Standardization. Subsequently, they held various administrative positions such as Head of LP3 Banyuwangi in 2014, Head of Academic and Student Affairs, and Head of General Administration. They have authored several textbooks in the field of aviation, particularly for Flight Operations programs, and are an active researcher with publications in both Google Scholar and Scopus indexed journals. Besides aviation, their areas of expertise also include management. Actively involved in various academic organizations, including serving as the Chairperson of the Technical Operations Expert Council at the Indonesian Aviation Transportation Professionals Forum (FORKOM PATI), and as the Chairperson of the Partnership Department at the Curug Alumni Association. They previously held the position of Chairperson of the Curug Alumni Pilots Association from 2019 to 2022.

Darmeli Nasution was born in Tapanuli Selatan, North Sumatra, Indonesia. She is the fourth child of H. Muhammad Jasen Nasution (deceased) and Hj. Siti Hasni Hasibuan (deceased). She has two children, Nurinziva Zahra Wardhani Pulungan and Ahmad Fakhrizal Pulungan, and is married to Wandan Pulungan. She completed her undergraduate studies in Management Informatics at Gunadarma University, Jakarta, in 1996 and continued her education with a Master's degree in Computer Science at the Postgraduate Program of Universitas Putra Indonesia (UPI-YPTK) Padang, West Sumatra, graduating in 2008. She earned her Doctorate in Computer Science from Universitas Sumatera Utara (USU) in 2024. She began her career as a lecturer in 2000 at the Faculty of Engineering, Universitas Pembangunan Panca Budi Medan and continues to be an active permanent lecturer in the Master's Program in Computer Science at the same university. She has published textbooks such as "Algorithms and Programming" in 2012, "Artificial Neural Networks (Prediction Algorithms & Implementation)" in 2020, and "Applications of Microcontrollers AT89S51 & ATMEGA 16 Using BASCOM," along with works on "Air Transportation and Regional Economic Development in Indonesia" and "Analysis of the Relationship Between Transportation and Standard of Living in Indonesia." She is also active in both national and international scientific publications, and several of her research works have been indexed in Google Scholar and Scopus. Her current research interests lie in Artificial Intelligence, specifically "Machine Learning" and "Deep Learning." She is an active member of various scientific organizations such as APTIKOM and the Association of Indonesian Engineers (PII). She can be contacted via email at: darmelinasution@gmail.com

Part III
Future Directions

Humanitarian Logistics and Lessons Learned from COVID-19

14

Sharina Osman

Abstract

The COVID-19 pandemic has presented unprecedented challenges to global humanitarian logistics, impacting supply chains, transportation networks, and emergency response operations. This comprehensive review examines the lessons learned from the pandemic in the context of humanitarian logistics, drawing from interdisciplinary literature in logistics, supply chain management, public health, and disaster management. By analysing key challenges, responses, and adaptations during the COVID-19 crisis, this paper identifies critical insights and best practices for enhancing the resilience and effectiveness of humanitarian logistics in future pandemics and emergencies.

Keywords

Humanitarian logistics · COVID-19 · Pandemic response · Supply chain management · Disaster management · Lessons learned

S. Osman (✉)
Business School, Universiti Kuala Lumpur, Kuala Lumpur, Malaysia
e-mail: sharina@unikl.edu.my

14.1 Introduction

The emergence of the COVID-19 pandemic has posed significant challenges to global humanitarian logistics, disrupting supply chains, straining transportation networks, and testing the resilience of emergency response systems (Sinha and Van de Ven 2020). The unprecedented scale and complexity of the pandemic have underscored the need for effective coordination, innovation, and collaboration in addressing humanitarian needs and delivering essential services to affected populations (Christopher et al. 2020).

Humanitarian logistics is a critical component of disaster response, involving the planning, implementation, and control of the movement and storage of goods, services, and information to meet the urgent needs of affected populations during disasters. It is a complex and multifaceted field that requires rapid decision-making, coordination among diverse stakeholders, and the flexibility to adapt to evolving situations on the ground.

The COVID-19 pandemic, with its rapid spread and global reach, has significantly challenged traditional humanitarian logistics models, exposing gaps in preparedness, response capabilities, and the integration of technology (Altay and Green 2006; Kovács and Spens 2022). Key issues included the inability to anticipate demand surges, the lack of stockpiling of essential supplies, and the inadequate integration of technology in logistics operations.

N. A. A. Rahman et al. (eds.), *Supply Chain in Humanitarian Operations*, Sustainable Development Goals Series,
https://doi.org/10.1007/978-981-96-4426-1_14

This literature review explores the current landscape of humanitarian logistics, with a focus on the lessons learned from the COVID-19 pandemic and the emerging trends and innovations shaping the future of the field.

14.2 Challenges in Humanitarian Logistics During Crisis

The efficient and effective delivery of aid during disasters is complicated by the numerous inherent challenges that humanitarian logistics encounters. Resource constraints, unpredictable demand, complex and frequently disrupted supply chains, and the necessity of coordination among various stakeholders, such as governments, non-governmental organisations (NGOs), and private sector partners, are among the challenges that must be addressed (Van Wassenhove 2006). Humanitarian logistics must operate in frequently turbulent and volatile environments, including conflict zones, areas affected by natural disasters, and, more recently, global health crises such as the COVID-19 pandemic, which further complicates these operations.

Supply Chain and Unpredictable Demand Disruptions

The unpredictability of demand for products and services is one of the primary challenges in humanitarian logistics. Disasters can result in abrupt increases in demand for essential items, including food, water, medical supplies, and shelter, which can overtax existing supply chains and result in shortages (Balcik et al. 2010). This challenge was exemplified by the COVID-19 pandemic, as the abrupt global demand for personal protective equipment (PPE), ventilators, and vaccines significantly exceeded the supply, resulting in critical shortages and delays (Queiroz et al. 2020). Lockdowns, border closures, and movement restrictions worsened supply chain disruptions, emphasising the necessity of more adaptable and resilient supply chain models in humanitarian logistics.

Funding Challenges and Resource Constraints

The capacity to respond to disasters effectively is frequently impeded by the limited resources and funding that humanitarian operations are often constrained by. The funding of humanitarian aid is typically reactive and fluctuates in accordance with the priorities of donors and the visibility of a crisis (Tomasini and Van Wassenhove 2009). This funding volatility can result in gaps in preparedness and response capabilities, as was demonstrated during the initial stages of the COVID-19 pandemic when numerous humanitarian organisations encountered difficulty in mobilising the requisite resources to expand operations (Heaslip et al. 2021). Additionally, the pandemic emphasised the necessity of pre-disaster funding and investment in preparedness to facilitate more efficient and effective responses.

Information Sharing and Coordination

Successful logistics operations necessitate effective coordination among humanitarian actors; however, this obstacle persists. The coordination of humanitarian logistics is frequently inefficient and complex due to the involvement of a diverse array of stakeholders, each with its own objectives, procedures, and capabilities (Pettit and Beresford 2009). Inefficiencies in resource allocation, voids in service delivery, and duplication of efforts can result from the absence of standardised processes and information-sharing mechanisms. The COVID-19 pandemic underscored the necessity of enhanced coordination and communication among stakeholders to guarantee that resources are deployed where they are most required (Heaslip et al. 2021).

14.3 Impact of COVID-19 on Humanitarian Logistics

Disruptions in the Supply Chain

The COVID-19 pandemic exposed the vulnerability of worldwide supply networks, affecting the accessibility and provision of vital commodities and services required for humanitarian relief operations (Siddiqui et al. 2020). The imposition of lockdowns, travel restrictions, and border closures has hindered the transportation of humanitarian assistance, medical supplies, and personnel, so worsening the already prevalent logistical difficulties in areas affected by the crisis (Ivanov 2020). An abrupt increase in the need for personal protective equipment (PPE), medical gadgets, and pharmaceuticals has placed significant pressure on supply networks, resulting in scarcities and delays in the acquisition and distribution processes (Pettit et al. 2022). For example, the sharp increase in demand for personal protective equipment (PPE) resulted in significant shortages, as prominent manufacturers were unable to cater to rapidly increasing output. The World Health Organisation (WHO) has documented substantial delays in the acquisition and dissemination of vital health supplies, which have had a disastrous impact on the response efforts in several low- and middle-income countries (WHO 2020).

The global scarcity of personal protective equipment (PPE) was apparent during the initial stages of the epidemic. Developed nations such as the United States and the United Kingdom faced significant scarcities, which led governments to investigate other sourcing strategies, such as domestic production and contributions from private enterprises. For example, in the United States, corporations like 3M and Honeywell expanded their manufacturing capabilities, while domestic producers shifted their focus to fabrication masks and face shields. A supply chain for personal protective equipment (PPE) was established by the UK government in collaboration with local businesses. This involved utilising the National Health Service (NHS) and private sector partnerships to procure and deliver necessary supplies (Berkley 2020).

In Malaysia, in response to significant shortages in PPE, the government undertook an investigation into alternate sourcing strategies, such as domestic production and contributions from private enterprises. For example, companies like Top Glove, the leading global producer of rubber gloves headquartered in Malaysia, extensively increased their production to satisfy the higher demand for gloves. Furthermore, local universities and research organisations joined together to manufacture face shields and other personal protective equipment (PPE) utilising 3D printing technology (Aziz et al. 2020).

Resource Allocation Challenges

The epidemic intensified the competition for few resources. Insufficient coordination and planning resulted in operational inefficiencies in the allocation and distribution of resources. The scarcity of ventilators and intensive care unit (ICU) equipment, especially during the initial phases of the epidemic, brought attention to these concerns (Oloruntoba & Gray 2006). In Italy and New York, hospitals experienced acute ventilator shortages, which overwhelmed healthcare facilities and placed significant burden on medical personnel (Hewitt et al. 2020). The state's healthcare system was inundated, with hospitals functioning at maximum capacity. The state's emergency management system confronted challenges in efficiently distributing ventilators and other essential equipment, resulting in the establishment of improvised facilities and the temporary transformation of convention centres into hospitals. The Jacob Javits Centre underwent a conversion into a temporary hospital with a capacity of over 2000 beds, which served to expand critical care services and relieve strain on current hospitals (Kang et al. 2020). Furthermore, the state employed the USNS Comfort, a navy hospital vessel, to offer supplementary healthcare services.

In Malaysia, the healthcare sector encountered difficulties in efficiently delivering ventilators and other essential equipment. The nation's emergency management system faced challenges in distributing available resources fairly among several states. To address this issue, the government formed partnerships with private sector entities to procure and disseminate ventilators. Specifically, the corporate sector, which includes corporations such as Petronas, provided financial support and resources to acquire ventilators and other medical equipment, therefore strengthening the healthcare system's ability to handle the increase in COVID-19 cases (The Star 2020).

Healthcare System Strain

Healthcare systems globally were stretched to their maximum capacity. The exponential growth in COVID-19 cases overwhelmed hospitals, resulting in shortages of medical staff, intensive care unit beds, and vital supplies. The exponential surge in COVID-19 patients in Italy resulted in a severe inadequacy of intensive care unit (ICU) beds, compelling hospitals to adopt triage procedures (Liu et al. 2020). The situation necessitated the utilisation of field hospitals and the adaptation of pre-existing facilities, such as sports halls and hotels, to specifically cater to Covid-19 patients (Liu et al. 2020).

Similarly, in New York, the healthcare system in New York faced significant challenges due to a sudden increase in cases, resulting in the establishment of temporary hospitals and the utilisation of improvised facilities to accommodate patients (Kang et al. 2020). The state transformed the Jacob Javits Centre and the USNS Comfort into temporary medical facilities, greatly improving the healthcare system's ability to manage the increased number of patients.

In Malaysia, the domestic healthcare system was inundated, with hospitals functioning at over their maximum capacity. The insufficient availability of intensive care unit (ICU) beds needed healthcare practitioners to establish triage procedures, which would prioritise treatment for patients according to their probability of survival. This emergency necessitated the utilisation of field hospitals and the adaptation of pre-existing facilities, such as sports halls and hotels, to house individuals afflicted with COVID-19. As an example, the Malaysian Armed Forces established a field hospital at the Malaysia Agro Exposition Park Serdang (MAEPS) to offer extra patient beds and medical services in order to relieve strain on current hospitals.

Societal and Economic Consequences

The epidemic precipitated profound economic recessions, therefore diminishing financial resources allocated to humanitarian relief and exacerbating poverty and displacement. World Bank's projection indicates that the world economy saw a contraction of 3.5% in 2020, resulting in the displacement of millions into poverty (World Bank 2020). In addition to their vulnerability, refugees and displaced people had other obstacles in obtaining healthcare and vital services (UNHCR 2020).

Within nations such as Syria and Yemen, the economic repercussions of COVID-19 intensified pre-existing humanitarian emergencies. The decrease in financial resources allocated to humanitarian assistance resulted in a deterioration of vital services such as healthcare, nutrition, and clean water, therefore intensifying the circumstances for fragile communities. In Yemen, the World Food Programme (WFP) has documented a 16% surge in the population experiencing severe food insecurity during the epidemic, resulting in a total of 16.2 million individuals. The rise in food insecurity was primarily caused by interruptions in supply networks, economic volatility, and the decrease in financial support for humanitarian assistance (UNHCR 2020). Moreover, refugees and undocumented migrants, notably those from Myanmar and Rohingya populations, encountered further obstacles in obtaining healthcare and vital services, underscoring the necessity for more comprehensive policy (UNHCR 2020).

The economic repercussions of COVID-19 also amplified pre-existing humanitarian concerns in Malaysia. The decrease in financial resources allocated to humanitarian assistance resulted in a deterioration of vital services such as healthcare and food. Specifically, the Malaysian government introduced the Prihatin Rakyat Economic Stimulus Package (PRIHATIN) in March 2020 to offer monetary aid to those and enterprises who have been impacted (MOF 2020). Encompassing RM 250 billion, the package sought to bolster individuals, enterprises, and the economy amidst the crisis (World Bank 2020; The Star 2020). The primary goals encompassed providing financial aid to individuals, offering savings on electricity, implementing loan moratoriums, and facilitating EPF withdrawals (New Straits Times 2020). Commercial enterprises were granted salary subsidies, micro-credit programmes, and tax exemptions. Allocation of healthcare funds was directed on enhancing the healthcare system and guaranteeing food security. Supplementary measures encompassed Bantuan Prihatin Nasional (BPN), a direct monetary assistance initiative targeting marginalised populations, as well as rent rebate. PRIPATIN played a pivotal role in achieving economic stability in Malaysia and effectively addressing urgent humanitarian requirements. Nevertheless, the situation also brought attention to weaknesses in Malaysia's economy and social structures, as the decrease in financial resources allocated to areas like healthcare and food security revealed the reliance of marginalised groups on continuous assistance from both the government and non-governmental entities (DOSM 2020).

14.4 Lessons Learned from the COVID-19 Pandemic

The COVID-19 epidemic has served as a clear and urgent reminder for the humanitarian logistics industry, exposing substantial deficiencies in readiness and ability to respond successfully. Salient insights gained from the epidemic encompass the necessity for improved foresight of increases in demand, the significance of accumulating vital resources, and the pivotal function of technology in augmenting logistical operations.

Forecasting Increases in Demand and Accumulating Inventory

The COVID-19 epidemic highlighted the urgent need for improved prediction of sharp increases in demand and the significance of keeping strategic reserves of vital resources. The abrupt surge in worldwide demand for personal protective equipment (PPE), medical supplies, and vaccinations revealed the weaknesses of just-in-time supply chain principles, emphasising the need for more robust inventory management approaches (Kovács and Sigala 2021). Consequently, humanitarian organisations are progressively adopting predictive analytics and demand forecasting to more accurately predict future requirements and guarantee the prompt availability of essential resources. The World Health Organisation (WHO) during the epidemic, created an innovative predictive analytics system to predict the demand for personal protective equipment (PPE) and vaccines in various region. By employing data-driven algorithms, the World Health Organisation (WHO) successfully determined the locations with the highest level of need and distributed resources appropriately. The implementation of these forecasting methods facilitated the World Health Organisation (WHO) in efficiently overseeing worldwide inventories, therefore guaranteeing the availability of supplies at the most critical times and locations.

A regional stockpile, such as the one established by the World Health Organisation (WHO) in Dubai, can strengthen the ability to address forthcoming health crises. The retention of these reserves guarantees the immediate availability of vital resources, hence diminishing reliance on worldwide supply networks. The World Health Organisation's centre in Dubai is furnished with medical supplies, vaccines, and personal protective equipment (PPE), prepared for swift distribution to areas requiring assistance. By implementing this approach, not only is the resilience of the supply chain improved, but also the response times are reduced, therefore assuring the timely delivery of relief to impacted regions (WHO 2020).

Using the National Crisis Preparedness and Response Centre (CPRC), the Malaysian government and healthcare agencies monitored real-time data on COVID-19 patients and forecasted future medical supply requirements. Through the use of predictive analytics, the Ministry of Health successfully predicted shortages in personal protective equipment (PPE) and medical equipment. This enabled the country to adequately manage strategic stocks and avoid crucial gaps in supply. The use of this pre-emptive strategy guaranteed that Malaysia's healthcare system was more prepared to manage increasing levels of demand during the epidemic.

In both the global and Malaysian contexts, the adoption of predictive analytics and more robust inventory management strategies has been crucial in mitigating supply chain weaknesses and enhancing readiness for future emergencies.

Augmenting the Resilience of Supply Chains

Following the COVID-19 epidemic, the establishment of robust supply chains has emerged as a crucial objective. In order to alleviate supply chain interruptions, numerous organisations transitioned to sourcing and producing locally. This entails the implementation of strategies such as diversifying supply sources, improving local production capabilities, and developing contingency plans to mitigate the consequences of future interruptions (Ivanov and Dolgui 2020). The epidemic underscored the benefits of using local resources and suppliers to decrease dependence on cross-border supply networks, which are susceptible to disruptions.

In response to worldwide disruptions in the supply chain, Japan initiated the Supply Chain Resilience Initiative (SCRI), with the aim of reducing production dependence on China. Japan made strategic investments in enhancing its domestic manufacturing capacities and expanding its supply sources throughout Southeast Asia. By broadening Japan's supply base and decreasing reliance on single-country dependence, this strategy effectively reduced risks.

In Malaysia, the government implemented the National Economic Recovery Plan (PENJANA) to stimulate local manufacturing by urging enterprises to embrace local suppliers and allocate resources to enhance domestic production capabilities. This was especially apparent in the healthcare industry, as companies such as Top Glove and Hartalega, the leading glove makers worldwide, increased both domestic and international manufacturing to suit worldwide needs. In response to the epidemic, Malaysia effectively mitigated its reliance on global supply chains for vital medical goods by enhancing domestic manufacturing.

The above examples demonstrate the role of both international and domestic initiatives aimed at enhancing local output and expanding supply networks in bolstering resilience during catastrophic events.

Utilising Technology to Enhance Logistics

The COVID-19 epidemic has greatly expedited the incorporation of digital technologies and innovation in the field of humanitarian logistics (Saha et al. 2020). In order to improve supply chain visibility and operational efficiency, organisations have progressively embraced sophisticated technologies such as data analytics, artificial intelligence, and blockchain. Advanced technologies such as remote monitoring, real-time tracking systems, automated inventory management, and digital platforms for coordinating logistics have played a crucial role in tackling the intricate and changing difficulties arising from the epidemic (Dubey et al. 2021). Advancements in technology have facilitated immediate and clear visibility and transparency in supply chain processes, leading to enhanced decision-making and more effective allocation of resources (Ponomarov and Holcomb 2009). Furthermore, the implementation of drones, robots, and autonomous vehicles for the last stage of delivery and the distribution of humanitarian assistance without physical contact have reduced the hazards faced by both frontline personnel and recipients (Ivanov 2020).

The WHO's adoption of the COVID-19 Supply Chain System served as a prime example of the application of blockchain technology in supply chain management. By improving transparency and traceability, this system enabled stakeholders to monitor the real-time movement of medical supplies, therefore minimising delays and guaranteeing the delivery of help to the intended recipients. Furthermore, the implementation of the blockchain technology facilitated the secure exchange of data between global health institutions, therefore mitigating the potential for fraudulent activities and improving the effectiveness of the supply chain (WHO 2020).

For the purpose of enhancing the transparency and efficiency of food distribution, especially in refugee camps, the World Food Programme (WFP) implemented blockchain technology. Blockchain-enabled platforms, such as the Building Blocks initiative, effectively facilitated the real-time monitoring of aid distribution, therefore enhancing coordination and mitigating the potential for fraud and wastage. This system optimised the supply chain, minimised dependence on intermediaries, and improved transparency in the allocation of food assistance.

In countries such as Rwanda and Ghana, unmanned aerial vehicles (UAVs) were used to transport vital medical supplies, including vaccines and blood, to isolated regions. Drones played a crucial role in the virus pandemic by facilitating the contactless distribution of essential health supplies, therefore reducing the potential for virus transmission and efficiently reaching far-flung areas. Business enterprises such as Zipline in Ghana significantly contributed to enabling these operations.

The International Red Cross (IRC) uses artificial intelligence (AI)-based predictive analytics to evaluate the prospective locations of future epidemics and the corresponding requirements for humanitarian assistance. Through the utilisation of real-time data, the IRC has the capability to forecast increases in demand for resources like food, medical supplies, and protective equipment. This enables a more proactive and effective deployment of resources across different locations.

Originally created to facilitate the identification of contacts and the surveillance of COVID-19 infection cases, Malaysia's MySejahtera App also made significant contributions to humanitarian logistics. The application's interface with the nation's vaccine distribution network offered real-time data on the availability and deployment locations of vaccines. The system's integration with supply chain platforms ensured the optimised distribution of vaccines and the minimisation of delays. Through the provision of real-time data on COVID-19 cases to authorities, the programme also enabled the effective administration of the allocation of medical supplies and resources. The application of blockchain technology in supply chain management, demonstrated by the collaboration between the Malaysian government and private technology firms, improved transparency and traceability, ensuring timely delivery of aid to its designated beneficiaries (Lee et al. 2020). MySejahtera is an exemplary demonstration of the integration of technology in logistics operations in Malaysia.

Additionally, the largest pharmaceutical business in Malaysia, Pharmaniaga, employed automated warehouse management systems to manage the storage and distribution of vaccines. Throughout the epidemic, the corporation depended on digital tools to monitor inventories in real time, guaranteeing the effective distribution of vaccines nationwide. The technology facilitated smooth, efficient collaboration among suppliers, health authorities, and distribution networks.

In response to the epidemic, Malaysia's Ministry of Health (MOH) also investigated the application of drones for the transportation of medical supplies and personal protective equipment (PPE) to remote regions. This programme successfully mitigated the risk of virus spread while ensuring prompt and effective delivery of vital assistance to rural populations.

These instances demonstrate the groundbreaking impact of digital technology on humanitarian logistics, namely in enhancing real-time monitoring, transparency, and the distribution of vital materials throughout the epidemic. In the future, the continued incorporation of technology into humanitarian logistics will be essential for improving the speed, precision, and overall efficiency of disaster response operations.

Enhancing Effective Coordination and Collaboration

The COVID-19 epidemic highlighted the crucial need of enhancing coordination and cooperation among humanitarian entities, international governments, and the commercial sector. Collaborative platforms and information-sharing systems played a crucial role in enabling the delivery of vaccines and medical supplies to people that are at risk (Kovács and Spens 2022). Multi-stakeholder collaborations and the establishment of standardised procedures are increasingly acknowledged as essential for improving coordination and minimising inefficiencies in humanitarian logistics.

The epidemic has highlighted the need of strong collaborations across many sectors, including governments, international organisations, non-governmental organisations (NGOs), and private sector entities, in order to allocate resources, exchange information, and synchronise response efforts (Pettit et al. 2022; Christopher et al. 2020). The Global Health Cluster and the Logistics Cluster were crucial in promoting coordination and information exchange among humanitarian stakeholders, therefore enhancing the comprehensiveness and efficiency of the crisis response (Tatham and Kovács 2020). The World Health Organisation (WHO) has been instrumental in its coordination endeavours, notably by implementing the COVID-19 Supply Chain System, which effectively optimised the allocation of essential resources (WHO 2020).

The efficacy of integrated processes in managing supply chain logistics during the epidemic was proved by case studies, such as the coordinated response in Bangladesh (Gow et al. 2020). Specifically, the government collaborated with groups like BRAC and the International Federation of Red Cross and Red Crescent Societies (IFRC) to create distribution networks that are based in the local communities. The established networks successfully sent food assistance, medical resources, and hygiene kits to marginalised communities in isolated regions, therefore improving the overall effectiveness of the response (Gow et al. 2020).

Collaborative efforts between the Malaysian government, international non-governmental organisations (NGOs), and local organisations were integral to the coordinated response to COVID-19 in Malaysia. The collaboration assured more efficient allocation of resources, expanded healthcare provisions, and refined community engagement initiatives. For example, the Malaysian government collaborated with organisations like as MERCY Malaysia and the Malaysian Red Crescent Society (MRCS) to create distribution networks that are based in the local community. The established networks successfully sent food assistance, medical resources, and hygiene kits to marginalised communities in isolated regions, therefore improving the overall effectiveness of the response (Gow et al. 2020).

Engagement and Empowerment of the Community

The COVID-19 epidemic has shown the immense significance of involving and empowering communities in the field of humanitarian logistics (Tatham and Kovács 2020). The pandemic response was significantly influenced by the essential contribution of local communities, who provided their expertise, resources, and networks to facilitate relief operations (Saha et al. 2020). By using participatory approaches like community-based logistics and social mobilisation, communities successfully cultivated trust, ownership, and resilience at the local level, thereby improving their capacity to regulate the consequences of the epidemic with greater efficiency (Ponomarov and Holcomb 2009).

The efficacy and scope of humanitarian endeavours were greatly enhanced by the empowering of local communities and volunteers. Community-based distribution networks and innovative local engagement initiatives enabled the efficient delivery of assistance to distant and inaccessible regions. For example, in remote areas of Africa, community-led action backed by local non-governmental organisations (NGOs) and international collaborators, successfully facilitated the delivery of necessary supplies and health information during the epidemic. This success showcased the effectiveness of localised, community-led strategies in humanitarian logistics (Munyua et al. 2020). In Kenya, the Kenya Red Cross Society (KRCS) partnered with community health volunteers were instrumental in the distribution of nutritious meals, clean water, and medical supplies to remote and disadvantaged communities. The implementation of these community-driven activities assured that assistance would be delivered to the most susceptible groups, therefore augmenting the efficiency of the response. The implementation of this grassroots strategy not only expanded the scope of the reaction but also cultivated a sense of community ownership and resilience (Munyua et al. 2020).

Similarly, community engagement had a crucial role in effectively reaching vulnerable communities in Malaysia during the epidemic. Local volunteers and grassroots organisations, such as NGO Mercy Malaysia, collaborated rigorously with government authorities to distribute food, medical supplies, and personal protective equipment (PPE) to remote and disadvantaged regions. The Food Bank Malaysia project and similar community-based programmes depended on local expertise and networks to pinpoint households in need, therefore guaranteeing prompt and efficient delivery of assistance to the most vulnerable populations. Participatory methods enhanced the agency of local communities and bolstered their ability to withstand the challenges posed by the epidemic.

Both on a global and local scale, the active involvement of local communities and volunteers played a crucial role in augmenting the scope and efficiency of humanitarian endeavours, especially in regions that posed challenges in terms of usual logistical routes.

14.5 Implications for Future Practice and Research

The insights gained during the COVID-19 pandemic have substantial ramifications for the future implementation and study of humanitarian logistics. Insights gained from the epidemic can be utilised by practitioners to improve the resilience, efficacy, and sustainability of humanitarian response operations. Facilitating the advancement of research and development in logistics technologies and solutions is crucial for ensuring future readiness. Sponsoring start-ups and technology-based humanitarian solutions can stimulate innovation in the field of logistics operations. Establishing strategic alliances with technology firms and academic institutions, helps expedite the advancement and use of state-of-the-art technologies in humanitarian logistics. Establishing thorough training programmes for logistics experts is essential for cultivating a highly competent staff. Allocating resources towards simulation-based training and scenario planning exercises can significantly improve the preparedness and response capacities of humanitarian organisations. Engaging in partnerships with academic institutions to create tailored programmes in humanitarian logistics can provide a consistent provision of professionals with the necessary expertise. For example, the curriculum incorporates high-fidelity simulations of catastrophic situations, enabling participants to formulate and evaluate their reaction tactics inside a regulated setting. The use of this practical training method guarantees that experts are adequately equipped to manage intricate logistical issues in actual emergencies.

Investments in digital infrastructure, capacity building, and innovation have the potential to enhance the resilience of supply chains and facilitate the development of more flexible and responsive humanitarian logistics systems. The development of policies and strategies by policymakers can effectively foster collaboration, creativity, and community involvement in the context of pandemic preparedness and response endeavours. To enhance theoretical understanding and practical expertise in the field, researchers can delve further into the dynamics and interconnections of crucial elements that influence humanitarian logistics during pandemics and other intricate catastrophes.

14.6 Conclusion

The COVID-19 pandemic has profoundly affected humanitarian logistics, challenging the robustness of global supply chains and emergency response mechanisms. The situation has underscored the necessity for innovative, flexible, and technologically advanced strategies in humanitarian logistics. It has revealed deficiencies in conventional systems, emphasising the demand for more robust, flexible, and technology-oriented solutions. By capitalising on the insights gained and adopting new trends, humanitarian agencies can improve their logistics proficiency and more effectively serve communities in subsequent emergencies.

References

Altay N, Green WG (2006) OR models in disaster operations management: survey, trends, and opportunities. Eur J Oper Res 175(1):475–493

Aziz AA, Othman NH, Razak NIA, Ibrahim F (2020) Risk management practices in humanitarian supply chains during the COVID-19 pandemic: a case study of Malaysia. J Humanitarian Logistics Supply Chain Manage.

Balcik B, Beamon BM, Krejci CC, Muramatsu KM, Ramirez M (2010) Coordination in humanitarian relief chains: practices, challenges, and opportunities. Int J Prod Econ 126(1):22–34

Bank Negara Malaysia (2020) COVID-19 economic relief measures. Bank Negara Malaysia. Retrieved from https://www.bnm.gov.my/

Berkley S (2020) COVID-19 needs a big science approach. Sci 367(6485):1407–1407

Christopher M, Cripps H, McCarthy L, Murphy S (2020) Building resilient humanitarian supply chains: lessons from the COVID-19 pandemic. Supply Chain Manage Rev

Department of Statistics Malaysia (2020) Special report on COVID-19 impact on the Malaysian economy. Department of Statistics Malaysia. Retrieved from https://www.dosm.gov.my/

Dubey R, Bryde DJ, Foropon C, Tabaklar T (2021) COVID-19 and supply chain resilience: a research agenda for the future. Ann Oper Res 309(1):135–150

Gow J et al (2020) Enhancing coordination in humanitarian logistics: the COVID-19 experience. J Humanitarian Logistics Supply Chain Manage 10(2):195–212

Heaslip G, Kovács G, Haavisto I (2021) A decade of humanitarian logistics research: a systematic review and future directions. J Humanitarian Logistics Supply Chain Manage 11(1):27–45

Hewitt C et al (2020) Challenges in ventilator allocation during the COVID-19 pandemic: lessons from the frontlines. Crit Care Med 48(5):e496–e502

Ivanov D (2020) Predicting the impacts of epidemic outbreaks on global supply chains: a simulation-based analysis on the coronavirus outbreak (COVID-19/SARS-CoV-2) case. Transp Res Part E: LogisticsTransp Rev 136:101922.

Ivanov D, Dolgui A (2020) A digital supply chain twin for managing the disruption risks and resilience in the era of Industry 4.0. Prod Plann Control 31(1):1–14

Kang SJ et al (2020) South Korea's response to the COVID-19 pandemic: rapid deployment of mobile testing units. J Glob Health 10(1):e010354

Kovács G, Sigala IF (2021) Lessons learned from humanitarian logistics to manage supply chain disruptions. J Supply Chain Manag 57(1):41–47

Kovács G, Spens KM (2022) Humanitarian logistics in the wake of COVID-19: advancing the science of global supply chain resilience. *Supply Chain Manage Rev 23*(3):12–19

Lee H et al (2020) Real-time data analytics in humanitarian logistics: enhancing decision making during COVID-19. J Supply Chain Manag 56(2):72–87

Liu Y et al (2020) The impact of COVID-19 on healthcare systems: lessons from Italy and New York. Lancet 395(10231):1683–1684

Ministry of Finance, Malaysia (2020) Prihatin Rakyat economic stimulus package: FAQs and breakdown. Ministry of Finance. Retrieved from https://www.mof.gov.my/

MOH (2020) Ministry of Health Malaysia COVID-19 report. Retrieved from [MOH website]

Munyua P et al (2020) Community-driven COVID-19 response in rural Africa: lessons learned. J Glob Health 10(1):010359
New Straits Times (2020) PRIHATIN package breakdown: RM250 billion economic lifeline. New Straits Times, March 27. Retrieved from https://www.nst.com.my/
Oloruntoba R, Gray R (2006) Humanitarian aid: an agile supply chain? Supply Chain Manage Int J 11(2):115–120
Pettit S, Beresford A (2009) Critical success factors in the context of humanitarian aid supply chains. Int J Phys Distrib Logist Manag 39:450–468
Pettit S, Wang Y, Beresford A (2022) The impact of digitalization on contemporary and future logistics. In The digital supply chain pp. 111–125. Elsevier
Ponomarov SY, Holcomb MC (2009) Understanding the concept of supply chain resilience. Int J Logistics Manage 20(1):124–143
Queiroz MM, Ivanov D, Dolgui A (2020) Resilient supply chains: research and future directions. Int J Logist Manag 31(3):620–640
Saha S, Kumar V, Gupta P (2020) Humanitarian supply chain in pandemic: a case of COVID-19. Int J Oper Prod Manag 40(4):554–578
Siddiqui MA, Wang G, Mannan S (2020) Critical role of emergency supply chains in the context of COVID-19. J Risk Res 23(5):729–745
Sinha K, Van de Ven A (2020) Crisis management and supply chain disruptions: learning from COVID-19. J Supply Chain Manag 56(2):111–127
Tatham P, Kovács G (2020) The role of humanitarian logistics during COVID-19: a review. Int J Log Res Appl 23(5):459–471
The Star (2020) Malaysia ventilator shortage crisis. Retrieved from [The Star website]
Tomasini RM, Wassenhove LN (2009) Logistics of humanitarian aid
UNHCR (2020) Impact of COVID-19 on refugees and migrants in Malaysia. Retrieved from [UNHCR website]
Wassenhove LV (2006) Humanitarian aid logistics: supply chain management in high gear. J Oper Res Soc 57:475–489
WHO (2020) Global humanitarian response plan COVID-19. Retrieved from [WHO website]
World Bank (2020) Global economic prospects: the impact of COVID-19 on the global economy. Retrieved from [World Bank website]

Dr. Sharina Osman is a senior lecturer in the Management & Entrepreneurship section and deputy dean of Student Development & Campus Lifestyle at Universiti Kuala Lumpur Business School, Malaysia. She holds a PhD in Management from the University of Exeter, United Kingdom, and both a Master's in Human Resource Management and a BBA (Honours) in Marketing from Universiti Utara Malaysia (UUM). With over 15 years of university-level teaching experience, Dr. Osman has a robust background in academia, research, and industry. Her research interests include Identity, Image and Reputation, Organisational Culture, Human Resource Development, Tourism Destination Identity & Branding, Participative Tourism Development, and e-Tourism. Dr. Osman has published widely in these fields and is an active member of several editorial advisory boards. She also serves as a keynote and guest speaker at numerous academic seminars. Beyond academia, Dr. Osman brings extensive industry experience in Human Resources, Marketing (specifically Image & Branding), and Tourism. Her expertise allows her to assist organisations in enhancing competency and capacity through training and consultation services. She is a strong advocate of experiential learning and has conducted various workshops for both government and private sectors locally and internationally.

GPSR Compliance

The European Union's (EU) General Product Safety Regulation (GPSR) is a set of rules that requires consumer products to be safe and our obligations to ensure this.

If you have any concerns about our products, you can contact us on ProductSafety@springernature.com

In case Publisher is established outside the EU, the EU authorized representative is:

Springer Nature Customer Service Center GmbH
Europaplatz 3
69115 Heidelberg, Germany

Batch number: 10584780

Printed by Printforce, the Netherlands